Wind Resistant Design of Bridges in Japan

Yozo Fujino • Kichiro Kimura • Hiroshi Tanaka

Wind Resistant Design of Bridges in Japan

Developments and Practices

Springer

Yozo Fujino
Professor
Department of Civil Engineering
The University of Tokyo
7-3-1 Hongo, Bunkyo-ku
Tokyo 113-8656
Japan
fujino@civil.t.u-tokyo.ac.jp

Kichiro Kimura
Professor
Department of Civil Engineering
Faculty of Science and Technology
Tokyo University of Science
kichiro@rs.tus.ac.jp

Hiroshi Tanaka
Professor Emeritus
University of Ottawa
K1N 6N5, Ottawa, ON
Canada
htanaka@uottawa.ca

ISBN 978-4-431-54045-8 e-ISBN 978-4-431-54046-5
DOI 10.1007/978-4-431-54046-5
Springer Tokyo Dordrecht Heidelberg London New York

Library of Congress Control Number: 2011945956

Printed on acid-free paper

Springer is part of Springer Science+Business Media (www.springer.com)

Preface

A large number of long-span bridges, in particular cable-supported bridges, have been built in Japan over the past 30 years. The number of cable-supported bridges may exceed 300. The Akashi-Kaikyo Bridge and Tatara Bridge are typical examples.

There are many technical challenges in constructing long-span bridges. Among those challenges, including materials, analysis, seismic action, and so on, wind actions are very important. Tremendous amounts of research and technical development have been accomplished in these areas.

Long-span bridges are light, flexible, and low-damped structures. Wind-induced static and dynamic behaviors govern the structural design. A rational wind-resistant design requires profound and comprehensive understanding of wind characteristics and performance of flexible bridges under wind excitation.

In Japan, research on the performance of suspension bridges under wind excitation started in the 1950s. A large wind tunnel for full models of suspension bridges was constructed at the University of Tokyo around 1960 and active research began being conducted. The first guideline on wind-resistant design for the Honshu-Shikoku Bridge Project was issued in the middle of the 1960s.

The behavior of a bridge under wind excitation is very complex because it is an outcome of the interaction between wind flow and bridge structure. Wind is light and extremely deformable. Hence, forces on girders, pylons, and cables induced by wind flows are different even if the geometries of these structural members are slightly different. The dynamic forces on them vary as they move under wind flows, creating the so-called motion-dependent forces that are extremely nonlinear. Theoretical treatment has certain limitations; thus an experimental approach using a wind tunnel becomes absolutely necessary.

In the 1970s, many wind tunnel facilities mainly for bridge structures were built by industries and universities in Japan. As a result, experimental basic and applied research on wind–bridge interaction was very active. In the long-span bridge projects, enormous wind tunnel tests were conducted in the search for the configuration of bridge girders and pylons. Control technologies such as tuned-mass damper (TMD) or active control schemes were applied during the construction or even on a permanent basis. Field measurement to confirm the validity of wind tunnel tests was also performed. Rain–wind vibration of stay cables was first observed on the Meiko-Nishi Bridge in 1984 during its erection.

The results of active research and development gathered during the wind-resistant design and the practices on the long-span bridges over the past 30 years have been of enormous importance. Some of the results and research activities were published in international journals and at conferences, but they are rather limited in number and many are unknown to research communities outside Japan.

Around the year 2000, we formed a small group within the Japan Society of Civil Engineers to collect information and data on the wind-resistant design of long-span bridges such as design guidelines, wind tunnel test results, and structural control applications. These data and information were compiled as a report that we believe will become a valuable asset in the wind-engineering and bridge-engineering communities.

Yozo Fujino edited the contents of this book as the chief editor, and we would also like to acknowledge those who made valuable contributions to the book.

Yoshiyuki Daito	Sumitomo Heavy Industries, Ltd. (Chap. 8, Appendix 2)
Koichiro Fumoto	Formerly of the Honshu-Shikoku Bridge Authority (Chaps. 1 and 2)
Nobuaki Furuya	National Defense Academy of Japan (Chaps. 3 and 4, Appendix 1)
Nobuyuki Hirahara	Honshu-Shikoku Bridge Expressway Company, Ltd. (Chaps. 1 and 2)
Akihiro Honda	Mitsubishi Heavy Industries, Ltd. (Chaps. 4, 6, and 9)
Hiroshi Katsuchi	Yokohama National University (Chaps. 3, 7, and 9)
Kichiro Kimura	Formerly of the Kyushu Institute of Technology (Chap. 6, Appendix 1)
Shigeki Kusuhara	Honshu-Shikoku Bridge Expressway Company, Ltd. (Chap. 3)
Kazutoshi Matsuda	Formerly of the IHI Corporation (Chaps. 6 and 8)
Katsumi Mikawa	Formerly of the Hitachi Zosen Corporation (Chap. 6)
Takuya Murakami	JFE Steel Corporation (Chap. 8)
Fumiaki Nagao	Tokushima University (Chaps. 11 and 12)
Hiroshi Sato	Formerly of the Public Works Research Institute (Chaps. 1 and 2)
Shuichi Suzuki	Formerly of the Honshu-Shikoku Bridge Authority (Chaps. 4 and 9)
Hiroshi Tanaka	Formerly of the University of Ottawa (Chap. 10)
Shinsuke Tokoro	Formerly of the Mitsubishi Heavy Industries, Ltd. (Chap. 4)
Tomomi Yagi	Kyoto University (Chap. 10)
Hitoshi Yamada	Yokohama National University (Chap. 6)
Hiroki Yamaguchi	Saitama University (Chap. 5)
Takeshi Yoshimura	Formerly of Kyushu Sangyo University (Chap. 8)
Fumitaka Yoshizumi	Formerly of Mitsui Engineering & Shipbuilding Co., Ltd. (Chap. 8)

Lastly, support from the Japan Steel Bridge Association is gratefully acknowledged.

Summer 2011

Yozo Fujino
Kichiro Kimura
Hiroshi Tanaka

Contents

Chapter 1
Wind Resistant Design Codes for Bridges in Japan

1.1 Wind Resistant Design Practice for Bridges in Japan

Wind resistant design practice for highway bridges in Japan is generally in accordance with the *Specifications for Highway Bridges* (SHB) [1], which applies to the bridges with the span length of less than 200 m. However, even for the bridges whose span length exceeds 200 m, the practice often is to consider SHB at any rate, with some modifications appropriate for each particular case. In particular, if the wind-induced dynamic behavior of the bridge is of any concern, the verification of safety against dynamic excitation is conducted by the use of the *Wind Resistant Design Manual for Highway Bridges* (WRDM), which provides with the supplementary information related to the wind-induced dynamic problems of bridges in general. In case of the *Honshu-Shikoku Bridge Project* (HSB), however, which involved many long-span cable-supported bridges, special provisions for wind resistant design, particularly applicable to this project, were established. Since then, these provisions have been also applied to some other long-span bridges, again with some modifications appropriate for each individual case. In the following sections, each of these design codes is briefly explained.

1.1.1 Specifications for Highway Bridges

The *Specifications for Highway Bridges* (SHB) [1] consists of five parts, of which Part 1 describes the general definitions and common rules, including the design loads such as the wind loading. Parts 2, 3, 4 and 5, on the other hand, present the regulations applicable to steel superstructures, concrete superstructures, substructures and seismic design rules, respectively. In terms of the design against wind, these specifications simply provide the definition of the static wind load, the way it is to be loaded, and the load factors considering the combinations with other loading. The detailed description of the latest version (2002) is given in Sect. 2.1.

Y. Fujino et al., *Wind Resistant Design of Bridges in Japan*,
DOI 10.1007/978-4-431-54046-5_1, © Springer 2012

1.1.2 Wind Related Codes for the Honshu-Shikoku Bridge Project

An intensive and extensive research and development for the HSB project started in the 1960s, keeping in mind a valuable lesson from the old Tacoma Narrows Bridge. From the very beginning of the project, it was clearly understood that the assurance of bridge safety against wind, not only under its static action but also under dynamic excitation, would have the utmost importance for the bridge design. It was thus decided that special provisions are to be developed for the bridge design against wind-induced dynamic effects, exclusively for this project, based upon the best knowledge available from various research activities and academic achievements.

The first product of these activities was summarized by the Technical Advisory Committee for Honshu-Shikoku Bridge Project in the Japan Society of Civil Engineers (JSCE), and published as the *Guide Specification for Wind Resistant Design of HSB* (*1964*) [2]. This *Guide Specification* was revised many times after that, as summarized in Table 3.1, and eventually crystallized as the *Wind Resistant Design Standard for HSB* (*1976*) [3]. As being shown in Table 3.1, the final design up to the Seto-Ohashi Bridges and the Ikuchi Bridge (please refer to Appendix 2 for the locations of the bridges), both in 1988, was based on this *Standard* (*1976*). As it was mentioned earlier, these *Guideline* and *Standards* were also used for other long-span suspension bridges and cable-stayed bridges outside the HSB project.

After the completion of the Seto-Ohashi Bridges, the new design codes specifically for the Akashi-Kaikyo Bridge and the bridges for the Onomichi-Imabari Route were established [4, 5]. By this time, the proposed span length for the Akashi Bridge was extended up to 2 km, and the Tatara Bridge was decided to be a cable-stayed structure. Wind statistics at the bridge sites were better established by then, and also there was a considerable progress in understanding of the bridge-wind interactions. The new design codes were made, incorporating these additional contributions. A newly built large wind tunnel was also made available by then for the aeroelastic full-bridge model tests in simulated atmospheric boundary layer wind. At the same time, the detailed computer simulations of the bridge dynamics were carried out. Experience in design and construction of the HSB project was thus hand-in-hand with the progressing research activities and full scale observations of already completed bridges. The technical development through this period is considered to be one of the greatest wind engineering achievements in Japan in the second half of the twentieth century, and it has been reflected on the *Wind Resistant Design Standard for HSB* (*2001*) [WRDSHSB(2001)] [6], which is introduced in Sect. 3.1.

Needless to say that the wind tunnel tests often consist of the most essential part of the wind resistant design of long-span bridges. The first experience of wind tunnel tests [7] for a long-span suspension bridge in Japan was on the Wakato Bridge, with its 367 m main span and was completed in 1962. The wind tunnel tests of this bridge were conducted in 1956–1958. As soon as the Wakato

Bridge was completed, the investigation toward the HSB project started [8]. For the HSB project, the importance of wind tunnel tests was recognized so much that its testing procedure has been also specified [9–12] to assure the reliability of the experimental results. Specific procedure for the wind tunnel tests has been separately established for the Akashi-Kaikyo Bridge [11].

1.1.3 Wind Resistant Design Manual for Highway Bridges

As the highway construction has been very active in Japan in recent years, many long-span bridges other than those of the HSB project have been also designed and built. Because of this reason, the *Wind Resistant Design Manual for Highway Bridges* was prepared in 1991 [13, 14] based on the accumulated knowledge as well as the experience acquired through the HSB project. The contents of this manual are explained in Sect. 2.2. The main characteristics of this manual are as follows:

1. The manual is designed in a way so that the wind resistant design could be conducted without necessarily carrying out wind tunnel tests. This practice was made possible by utilizing the data obtained from the previously conducted wind tunnel tests.
2. The manual provides with the information regarding the influence of wind turbulence on the wind-induced bridge behavior.

1.2 Basic Philosophy of Wind Resistant Design in Japanese Practice

1.2.1 Design Wind Load and Gust Response

For the design of ordinary highway bridges, wind is considered as a statically applied load. The wind load per unit length of the plate-girder bridge, for example, defined in the SHB [1] is given in Table 2.1.

The wind load given in Table 2.1 is defined based on the design wind pressure given by $p = \rho U_d^2/2{\cdot}C_D G$; where $U_d = 40\,\mathrm{m/s}$ is the design wind speed considered as a 10 min average, in which various topographical conditions are assumed to be already counted for; C_D is the drag force coefficient which depends on the ratio of deck width to deck depth (B/D); and $G = 1.9$ is the gust response factor. Considering the fact that the wind load is not usually a decisively significant load for the bridge design, its formulation is very much simplified.

However, for the long-span cable-supported bridges, for which a substantial dynamic amplification of the lateral bending is eminent, the concept of an equivalent static load is introduced. The equivalent load is defined as a uniformly distributed static wind load that would produce the maximum internal reactions of the bridge, equivalent to the internal reactions under the wind-induced action including the dynamic gust. Evaluation of the gust response in the *Wind Resistant Design Standard for HSB* is based on the method proposed by Davenport [15], in which the gust-induced bridge response is estimated using a frequency domain analysis from the power spectral density of wind velocity fluctuation, the mechanical admittance of the bridge, and the aerodynamic admittance representing the dynamic interaction of wind and structure. After a few revisions, the design wind load per unit length of the suspended structure for the Akashi-Kaikyo Bridge was defined as follows:

$$P = \mu_2 \frac{\rho U_z^2}{2} C_D A \tag{1.1}$$

$$U_z = \mu_1 U_{10} \tag{1.2}$$

where μ_1 is a modification factor depending on the elevation of the structure above the ground level; U_z is the design wind speed, considered as a 10 min average at the reference elevation of the structure; and μ_2 is a modification factor to include the dynamic amplification effect, depending upon the characteristics of both wind and structure.

1.2.2 Design Consideration of Wind-Induced Vibration

Although the wind load thus defined already includes the dynamic effects of wind gust, the safety of a bridge and its structural members against wind-induced vibrations such as flutter or vortex-induced vibrations would require a separate attention for some bridges. The 1964 edition of the *Wind Resistant Design Standard for HSB* [3] was probably the first design provision ever in the world that requested such examinations either by analysis or by wind tunnel tests. Particularly, for the aerodynamically self-excited vibration, which would lead the structure to a catastrophic destruction, this code requested that the critical wind speed to observe such behavior has to be 1.2 times the design wind speed or higher, with additional consideration of the effect of wind speed fluctuation; i.e.,

$$V_{cr} \geq 1.2 \cdot \mu_F \cdot U_z \tag{1.3}$$

where V_{cr} is the critical wind speed where the self-excited vibration is observed; and μ_F is a modification factor considering the wind speed fluctuation as being introduced in Sect. 3.1.3.

In order to examine the safety of the bridge against wind-induced vibration, the *Wind Resistant Design Manual for Highway Bridges* [13, 16] suggests the following steps in the procedure:

1. Depending on the structural types and characteristics of the bridge and wind conditions, some specific categories of dynamic response should be considered.
2. A preliminary estimate of possible dynamic behavior is to be made using approximate methods and the safety and serviceability of the bridge are to be considered.
3. If any dynamic problems are anticipated, the following two consequences should be considered: (1) to conduct a more precise study to make prediction of dynamic bridge behavior, either based on the past experience of similar bridges or by carrying out wind tunnel tests for this particular bridge; and (2) to redesign the bridge if the problems persist.

1.2.3 Wind Resistant Design of the Akashi-Kaikyo Bridge

The Akashi-Kaikyo Bridge, with its main span of nearly 2 km, was a bridge of unprecedented character in many ways. Because of this reason, a wide range of aerodynamic investigation was carried out for this bridge, including both analytical and experimental studies, and the *Wind Resistant Design Specification for Akashi-Kaikyo Bridge* [11] was established as a particular design provision for this bridge. This specification has the following characteristics:

1. The gust response analysis is to be carried out not only in lateral bending behavior but also in vertical bending and torsional behavior of the bridge girder.
2. Vortex-induced vibration is to be examined in wind tunnel not only in smooth wind but also by considering the natural wind turbulence.
3. The final safety against aerodynamic self-excited vibration is to be examined by a 3D full-bridge model test.
4. Characteristics of natural wind turbulence at the bridge site were well compiled to determine the basic wind speed for design.
5. The analytical methods to calculate the gust response and wind tunnel test methods including the natural wind simulation were included in the specification.

The necessity to carry out the aeroelastic full-bridge model tests was eminent. It was also recognized that the linear scale of 1:100 or more to construct the bridge model was deemed desirable to ensure the accuracy of the model tests, which would require a width of 40 m for the wind tunnel test section. Such a large facility was not available in Japan at that time. A new wind tunnel with the test section of 41 m wide, 4 m high and 30 m long as shown in Fig. 1.1 was constructed in the campus of the Public Works Research Institute, the Ministry of Construction, as a collaborative research project with the Honshu-Shikoku Bridge Authority. The facility was large enough to accommodate a model of the Akashi Bridge with a desirable scale

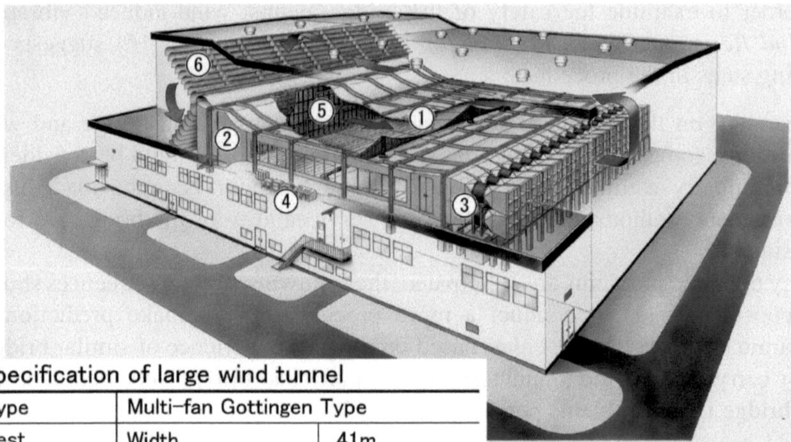

Specification of large wind tunnel

Type	Multi-fan Gottingen Type	
Test section	Width Height Length	41m 4m 30m
Fans	Diameter Number Wind Volume Rotation speed	1.8m 36 3,280m³/min 37.5–900rpm
Air flow	Wind speed Turbulent intensity	0.5–12m/s 1.0% or less

Fig. 1.1 Newly built boundary layer wind tunnel in Tsukuba

and test it in a turbulent boundary layer flow as a simulated natural wind at the bridge site. It was the world largest facility of this kind for civil engineering purposes. This boundary layer wind tunnel was fully utilized in later days for the aerodynamic study of the Akashi-Kaikyo Bridge, the Kurushima-Kaikyo Bridges as well as the Tatara Bridge, then the world's longest span cable-stayed bridge [17–20].

References

1. Japan Road Association (2002) Specifications for highway bridges (in Japanese). Maruzen, Tokyo
2. Japan Society of Civil Engineers (1964) Guide specification for wind resistant design of Honshu-Shikoku Bridges, Technical advisory committee for the Honshu-Shikoku bridge project (in Japanese). JSCE, Tokyo
3. Honshu-Shikoku Bridge Authority (1976) Wind resistant design standard for Honshu-Shikoku bridges (in Japanese). HSBA, Kobe
4. Honshu-Shikoku Bridge Authority (1990) Wind resistant design specification for Akashi-Kaikyo bridge (in Japanese). HSBA, Kobe
5. Honshu-Shikoku Bridge Authority (1994) Wind resistant design standard for Onomichi-Imabari route (in Japanese). HSBA, Kobe
6. Honshu-Shikoku Bridge Authority (2001) Wind resistant design standard for Honshu-Shikoku bridges (in Japanese). HSBA, Kobe

7. Hirai A et al (1960) An experimental study on the wind resistance of Wakato bridge (in Japanese). Technical Report of Bridge Engineering Laboratory, University of Tokyo, pp 1–201
8. Hirai A (1967) Wind resistance of suspension bridges (in Japanese). In: Steel bridges, Part III. Giho-Do, Tokyo, pp 454–579, Chapter 10
9. Japan Society of Civil Engineers (1973) Wind tunnel test specification for Honshu-Shikoku bridges (in Japanese). Wind resistant design committee for the Honshu-Shikoku bridges
10. Honshu-Shikoku Bridge Authority (1980) Wind tunnel test specification for Honshu-Shikoku bridges (in Japanese). HSBA, Kobe
11. Honshu-Shikoku Bridge Authority (1990) Wind tunnel test specification for Akashi-Kaikyo bridge (in Japanese). HSBA, Kobe
12. Honshu-Shikoku Bridge Authority (2001) Wind tunnel test specification for Honshu-Shikoku bridges (in Japanese)
13. Japan Road Association (1991) Wind resistant design manual for highway bridges (in Japanese). Maruzen, Tokyo
14. Sato H (1993) On the wind resistant design manual for highway bridges in Japan. In: Proceedings of the first US-Japan workshop on the advanced technology in highway engineering, pp 177–186
15. Davenport AG (1962) Buffeting of a suspension bridge by stormy winds. J Struct Div ASCE 88(ST3):233–268
16. Japan Road Association (2007) Wind resistant design manual for highway bridges, revised version in 2007 (in Japanese). Maruzen, Tokyo
17. Katsuchi H, Miyata T, Kitagawa M, Sato H, Hikami Y (1994) Study on characteristics of coupled-flutter in Akashi Kaikyo bridge's full model wind tunnel test (in Japanese). In: Proceedings of the 13th national symposium on wind engineering, pp 383–388
18. Miyata T, Sato H, Toriumi R, Kitagawa M, Katsuchi M (1995) Full model wind tunnel study on the Akashi Kaikyo bridge. In: Proceedings of the 9th international conference on wind engineering, New Delhi, pp 793–814
19. Miyata T, Tada K, Sato H, Katsuchi H, Hikami Y (1994) New findings of coupled-flutter in full model wind tunnel tests on the Akashi Kaikyo bridge. In: Proceedings of the cable-stayed and suspension bridges, Deauville, pp 163–170
20. Miyata T, Kitagawa M, Yamada H, Kanazaki T, Toriumi R (1994) Design considerations on 3-dimensional gust response in the Akashi Kaikyo bridge. In: Proceedings of the cable-stayed and suspension bridges, Deauville, pp 171–178

Chapter 2
Wind Resistant Design of Highway Bridges in General

2.1 Wind Load Specification for Highway Bridges

2.1.1 Summary

The *Specifications for Highway Bridges* (SHB) [1] is the most basic highway bridge design code in Japan. It applies to the bridges with the span length of less than 200 m (inclusive). Part 1 of the 5 volume specifications specifies the design wind load and the way it should be applied.

2.1.2 Wind Load and Its Application

The wind load for the bridge superstructure is horizontally distributed and applied normal to the longitudinal bridge axis. The design wind load is to be applied in a way so that the worst condition is to be considered for the design of each member. However, in case the bridge has a noise barrier installed, the wind loading conditions could be relaxed, considering the wind environment at the site and also the structural details of the noise barrier.

The reduction of wind load here is considered partly because the noise barriers would not be required at the first place, unless the surrounding area is heavily built-up. If indeed the area has a built-up topography, it would be reasonable to expect some reduction of the mean wind speed. It is also considered that the noise barriers are often designed in such a way that the wind load on the barriers will not be fully transmitted to the bridge structure by, for example, opening up holes, in case the wind speed is high.

Y. Fujino et al., *Wind Resistant Design of Bridges in Japan*,
DOI 10.1007/978-4-431-54046-5_2, © Springer 2012

Table 2.1 Wind load for plate girders (kN/m)

Section profile	Wind load
$1 \leq B/D < 8$	$\{4.0 - 0.2\,(B/D)\}D \geq 6.0$
$8 \leq B/D$	$2.4\,D \geq 6.0$

B = Width of the bridge (m), see Fig. 2.1
D = Exposed depth of the bridge (m)

Table 2.2 Wind load for two-plane trusses (kN/m^2)

Trusses	With live load	$1.25/\sqrt{\phi}$
	Without live load	$2.5/\sqrt{\phi}$
Floor system	With live load	1.5
	Without live load	3.0

ϕ = Solidity ratio of the truss (projected area to the outer area) and $0.1 \leq \phi \leq 0.6$

Table 2.3 Wind load for standard two-plane trusses (kN/m)

Chord member		Wind load
Loaded chord	With live load	$1.5 + 1.5D + 1.25\sqrt{\lambda h} \geq 6.0$
	Without live load	$3.0D + 2.5\sqrt{\lambda h} \geq 6.0$
Unloaded chord	With live load	$1.25\sqrt{\lambda h} \geq 3.0$
	Without live load	$2.5\sqrt{\lambda h} \geq 3.0$

D = Depth of the floor system excluding the overlapping area with the chord members (m)
λ = Distance between the centerlines of the upper and lower chords (m)
h = Depth of each chord member (m), and $7 \leq \lambda/D \leq 40$

2.1.2.1 Plate Girders

Wind load for a plate girder bridge per its unit length is to be in accordance with Table 2.1.

2.1.2.2 Truss Bridges Composed of Two Main Frames

Wind load for the two-plane truss bridges is to be in accordance with Table 2.2, which defines the wind-induced pressure for a vertical projected area of the windward side. However, for a standard two-plane truss, to simplify the calculation, the wind load defined in Table 2.3 can be applied per unit length of the windward truss chord members.

2.1.2.3 Bridges of Other Types

Wind load for the bridges of other structural types than defined above is to be decided by considering the rules of Sects. 2.1.2.1 and 2.1.2.2.

Table 2.4 Wind load for the structural members (kN/m^2)

Cross-section of the member		Wind load	
		Windward members	Leeward members
Circular	With live load	0.75	0.75
	Without live load	1.5	1.5
Rectangular	With live load	1.5	0.75
	Without live load	3.0	1.5

Table 2.5 Wind load for substructures (kN/m^2)

Cross sectional shape of the body		Wind load
Circular or oval	With live load	0.75
	Without live load	1.5
Square or rectangular	With live load	1.5
	Without live load	3.0

For individual structural members that cannot be categorized in Sect. 2.1.2.1 or 2.1.2.2, Table 2.4 can be applied, depending upon their cross-sectional shapes. Wind load acting on the live load shall be 1.5 kN/m in magnitude and is to be applied at the elevation of 1.5 m above the upper surface of the bridge floor.

2.1.2.4 Bridges in Parallel

When two or more steel girder bridges are in closely parallel configuration, the wind load specified in Table 2.1 shall be duly modified considering the aerodynamic interaction between them.

2.1.2.5 Wind Load for Substructures

Wind load acting directly to substructures is to be a horizontal load in the bridge axis direction or in the perpendicular direction, but consideration of simultaneous loading for both directions is not required. The wind load intensity on the windward vertical effective projected area is to be in accordance with Table 2.5.

2.2 Wind Resistant Design Manual for Highway Bridges

2.2.1 Outline of the Manual

The *Wind Resistant Design Manual for Highway Bridges* (WRDM) [2] is supplementary to the *Specifications for Highway Bridges* (SHB) with its particular attention focused upon dynamic response of bridges. As already mentioned in Sect. 1.1.3, WRDM has been based on the accumulated knowledge and experience acquired through Honshu-Shikoku Bridge projects as well as many long-span bridge constructions in Japan, and WRDM intends to keep the wind resistant design on somewhat conservative side without conducting wind tunnel tests for individual bridges. Though the dynamic amplification of wind response due to gust has been already taken into consideration in SHB by the definition of design wind load (Sect. 2.1), wind-induced vibration such as vortex-induced vibration and/or self-excited divergent motions need to be considered separately. WRDM is to be basically applied for a bridge having the span length up to 200 m because SHB has aimed at such span range, but WRDM has also been often consulted in design of longer span bridges.

2.2.1.1 Terms and Notations

Main terms and notations used in the WRDM are as follows:

- Maximum span length, L (m): Maximum longitudinal length of a bridge between two of its supports
- Deck height, z (m): Height of the road surface at the middle of center span of the bridge measured from the water or ground surface
- Width, B (m): Transverse dimension of the bridge (Fig. 2.1)
- Exposed depth, D (m): Vertical dimension of the bridge deck including girder, floor and handrails
- Design wind speed, U_d (m/s): Wind speed for determination of the design wind load

2.2.1.2 Principle of Wind Resistant Design

Dynamic wind resistant design is usually carried out in the following steps:

1. Estimation of the wind characteristics at the site
2. Assessment of dynamic behavior of the bridge deck and/or very slender members of the bridge
3. Investigation toward the safety during erection

As for the estimation of wind characteristics at the site, the design wind speed and turbulence intensity are estimated in accordance with Sect. 2.2.2. When the dynamic design consideration is deemed necessary based on Sect. 2.2.3.1, the formulae specified in Sect. 2.2.3.3 can be employed for verifying the proposed bridge deck section against individual dynamic behavior. If the results are unsatisfactory, more detailed investigation is required in accordance with Sect. 2.2.3.4. Dynamic design of very slender members and consideration of safety during erection are referred to Sects. 2.2.4 and 2.2.5, respectively.

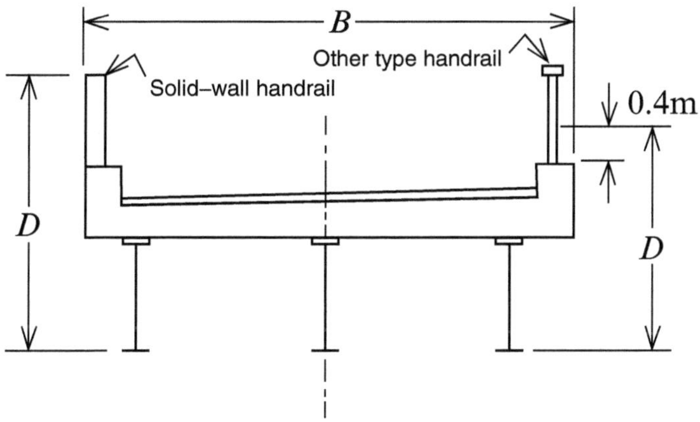

Fig. 2.1 Definition of *B* and *D*

2.2.2 *Wind Characteristics*

Design wind speed and turbulence intensity are estimated through the following procedure. Further discussion on this issue is referred to Sect. 4.3.

2.2.2.1 Design Wind Speed

The design wind speed U_d (m/s) is decided from the basic wind speed U_{10} and a modification factor E_1, which accounts for the influences of the height of the structure and the ground surface roughness of the area.

(a) *Basic wind speed* U_{10} (m/s) can be determined from Fig. 2.2, in which the values are normalized for the ground surface roughness of category II. The probability of not exceeding the given speed in 50 years is approximately 0.6.
(b) *Ground surface roughness* is to be identified as one of four categories given in Table 2.6. It is decided as an averaged terrain condition within the distance of approximately 100 times the bridge deck height z, but not less than 500 m. The area of this averaging is to be considered as shown in Fig. 2.3. When the bridge crosses a river, which is wider than 100 m, the roughness category is to be reduced by one rank compared to the case without the river.
(c) *Modification factor*, E_1 is decided by Table 2.7.
(d) *Design wind speed*, U_d (m/s) is defined as follows:

$$U_d = E_1 \cdot U_{10} \tag{2.1}$$

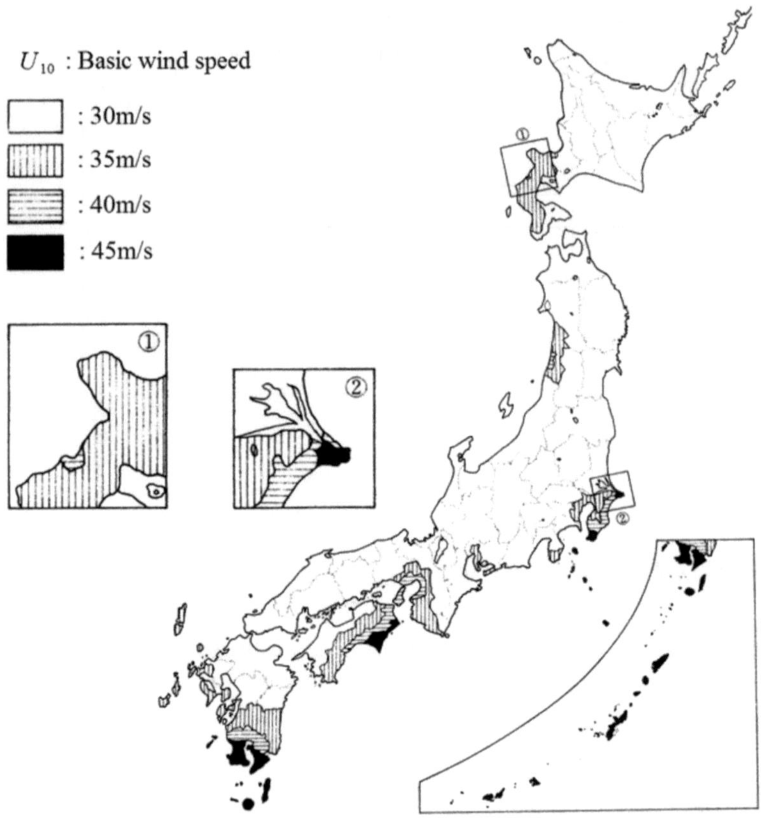

Fig. 2.2 Basic wind speed map of Japan [2]

Table 2.6 Category of ground surface roughness

Category	Surface condition	Surface condition note
I	Open sea, coastal area	Small roughness
II	Farmland, country side, open land	↑
	Area with scattered trees or low-rise buildings	
III	Area with dense trees of low-rise buildings	
	Area with scattered medium-rise buildings	↓
	Grand hills	
IV	Area with scattered high-rise buildings	High roughness
	Rugged hills	

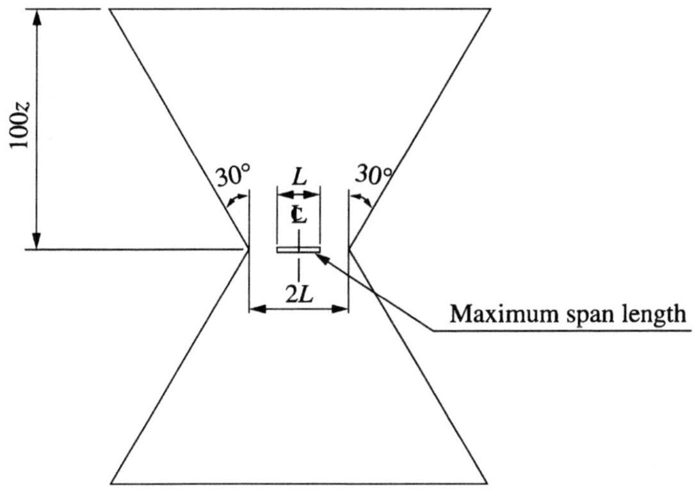

Note: z is the deck height specified in Sect. 2.2.1 (1)

Fig. 2.3 Area to be considered for an averaged terrain

Table 2.7 Modification factor E_1

Height (m) \ Surface roughness	I	II	III	IV
$z \leqq 5$	1.11	1.00	0.83	0.77
$5 < z \leqq 10$	1.16	1.00	0.83	0.77
$10 < z \leqq 15$	1.24	1.04	0.83	0.77
$15 < z \leqq 20$	1.29	1.09	0.85	0.77
$20 < z \leqq 25$	1.33	1.14	0.90	0.77
$25 < z \leqq 30$	1.36	1.18	0.94	0.77
$30 < z \leqq 35$	1.39	1.21	0.98	0.79
$35 < z \leqq 40$	1.41	1.24	1.01	0.82
$40 < z \leqq 45$	1.43	1.26	1.04	0.85
$45 < z \leqq 50$	1.45	1.28	1.07	0.88
$50 < z \leqq 60$	1.47	1.31	1.11	0.92
$60 < z \leqq 70$	1.50	1.35	1.15	0.96
$70 < z \leqq 80$	1.53	1.38	1.18	1.00
$80 < z \leqq 90$	1.55	1.41	1.22	1.04
$90 < z \leqq 100$	1.57	1.43	1.25	1.08

2.2.2.2 Intensity of Turbulence

The intensity of turbulence I_u is assumed in accordance with Table 2.8.

Table 2.8 Intensity of turbulence

Height (m) \ Surface roughness	I	II	III	IV
$z \leqq 10$	0.15	0.19	0.25	0.29
$10 < z \leqq 20$	0.14	0.17	0.25	0.29
$20 < z \leqq 30$	0.13	0.16	0.23	0.29
$30 < z \leqq 40$	0.12	0.15	0.21	0.28
$40 < z \leqq 50$	0.12	0.15	0.20	0.26
$50 < z \leqq 70$	0.11	0.14	0.18	0.24
$70 < z \leqq 100$	0.11	0.13	0.17	0.22

Table 2.9 Conditions where the wind-induced vibrations are possible

Type of bridges			Conditions	Possible phenomena
Suspension and/or cable-stayed bridges	Truss girders		$L \cdot U_d/B > 350$	Divergent vibration in torsion
	Solid web girders	Open cross section	$L \cdot U_d/B > 350$	Divergent vibration in torsion
			$L \cdot U_d/B > 330$ and $B/d < 5$ and $I_u < 0.15$ and steel bridge	Divergent vibration in bending
			$L \cdot U_d/B > 200$ and $I_u < 0.20$	Vortex-induced vibration
		Closed cross section	$L \cdot U_d/B > 520$	Divergent vibration in torsion
			$L \cdot U_d/B > 330$ and $B/d < 5$ and $I_u < 0.15$ and steel bridge	Divergent vibration in bending
			$L \cdot U_d/B > 200$ and $I_u < 0.20$	Vortex-induced vibration
Steel girder bridges	Plate or box girder		$L \cdot U_d/B > 330$ and $B/d < 5$ and $I_u < 0.15$	Divergent vibration in bending
			$L \cdot U_d/B > 200$ and $I_u < 0.20$	Vortex-induced vibration in bending

2.2.3 *Design of Bridge Deck*

2.2.3.1 Necessity of Dynamic Wind Resistant Design

Table 2.9 shows the conditions when the dynamic wind resistant design is deemed necessary.

Table 2.10 Coefficient E_{rl}

Surface roughness	I	II	III	IV
E_{rl}	1.10	1.15	1.20	1.25

Table 2.11 Structural damping of bridges (logarithmic decrement)

Type of bridges	Logarithmic decrement δ
Suspension bridge	
With truss girders	0.03
With solid-web girders	0.02
Cable-stayed bridge	
With truss girders	0.03
With solid-web girders	0.02
Steel girder bridges	$\frac{0.75}{\sqrt{L}}$, but $\delta \geq 0.04$

L: maximum span length (m)

2.2.3.2 Terms Required for the Design Calculation

(a) *Reference wind speed* U_r (m/s) is defined as follows:

For the instability in torsion (flutter)

$$U_{rf} = 1.2 \cdot E_{rl} \cdot U_d \tag{2.2}$$

For the instability in bending (galloping)

$$U_{rg} = 1.2 \cdot U_d \tag{2.3}$$

where, the coefficient E_{rl} is given in Table 2.10.

(b) *Structural damping* in logarithmic decrement δ is to be assumed as given in Table 2.11.

(c) *Natural frequencies* f (Hz) used for the safety assessment of the divergent type vibration in torsion or bending should be decided by eigenvalue analysis. For the assessment of the vortex-induced vibration, they could be assumed based on some empirical formulae for a quick examination, in the absence of properly calculated data.

(d) *Mass* m (kg/m) *and polar mass moment of inertia* I_p (kg.m^2/m) *per unit length of bridge deck* should be calculated based on the design. However, for the polar mass moment of inertia, the following formula may be used for simplicity:

$$I_P = (0.3B)^2 \cdot m \tag{2.4}$$

2.2.3.3 Assessment of Dynamic Safety

When the examination of dynamic safety is deemed necessary from Sect. 2.2.3.1, its assessment is to be carried out by using the following formulae:

(a) *Instability in torsion*

$$U_{cf} = 2.5 f_\theta \cdot B > U_{rf} \tag{2.5}$$

U_{cf}, U_{rf} and f_θ are the critical and reference wind speeds, and torsional natural frequency, respectively.

(b) *Instability in vertical bending*

$$U_{cg} = 8 f_h \cdot B > U_{rg} \tag{2.6a}$$

for almost horizontal wind

$$U_{cg} = 4 f_h \cdot B > U_{rg} \tag{2.6b}$$

for the wind with upward inclination angle

U_{cg}, U_{rg} and f_h are the critical and reference wind speed, and bending natural frequency, respectively.

(c) *Vortex-induced vibration in vertical bending*

$$U_{cvh} = 2.0 f_h \cdot B > U_d \tag{2.7}$$

and the maximum amplitude

$$h_c = \frac{E_h \cdot E_{th}}{m\delta} \rho B^3 < h_a \tag{2.8}$$

where, U_d = the design wind speed, and $h_a = 0.04/f_h$ = the allowable amplitude. Also, ρ is air density, E_{th} is defined in Table 2.12; and

$$E_h = \frac{0.065 \beta_{ds}}{B/D} \tag{2.9}$$

where β_{ds} is a coefficient specified for the deck configuration:

$$\beta_{ds} = \begin{cases} 2 & \text{when the length of bracket is less than } D/4 \text{ and web is vertical} \\ 1 & \text{cases other than above} \end{cases} \tag{2.10}$$

Table 2.12 Coefficient E_{th}

<table>
<tr><th rowspan="2">B/d \ I_u</th><th colspan="10">Cross-section other than hexagonal shape</th><th rowspan="2">Hexa-gon</th></tr>
<tr><th>0.11</th><th>0.12</th><th>0.13</th><th>0.14</th><th>0.15</th><th>0.16</th><th>0.17</th><th>0.18</th><th>0.19</th><th>0.20</th></tr>
<tr><td>$2 < B/d \leqq 3$</td><td>0.7</td><td></td><td colspan="2"></td><td>0.5</td><td colspan="2">0.4</td><td>0.3</td><td>0.2</td><td>0.1</td><td rowspan="9">1</td></tr>
<tr><td>$3 < B/d \leqq 4$</td><td></td><td rowspan="2">0.6</td><td></td><td>0.5</td><td>0.4</td><td>0.3</td><td>0.2</td><td>0.1</td><td></td><td></td></tr>
<tr><td>$4 < B/d \leqq 5$</td><td rowspan="3"></td><td rowspan="2"></td><td rowspan="2">0.4</td><td rowspan="2">0.3</td><td rowspan="2">0.2</td><td>0.1</td><td></td><td></td><td></td></tr>
<tr><td>$5 < B/d \leqq 6$</td><td rowspan="4">0.5</td><td></td><td></td><td></td><td></td></tr>
<tr><td>$6 < B/d \leqq 7$</td><td rowspan="2">0.4</td><td rowspan="2">0.3</td><td>0.2</td><td>0.1</td><td></td><td></td><td></td><td></td></tr>
<tr><td>$7 < B/d \leqq 8$</td><td rowspan="4"></td><td rowspan="2">0.1</td><td></td><td></td><td></td><td></td><td></td></tr>
<tr><td>$8 < B/d \leqq 9$</td><td rowspan="2">0.3</td><td rowspan="2">0.2</td><td></td><td></td><td></td><td></td><td></td></tr>
<tr><td>$9 < B/d \leqq 10$</td><td rowspan="2">0.4</td><td rowspan="2">0.1</td><td></td><td></td><td>0</td><td></td><td></td></tr>
<tr><td>$10 < B/d \leqq 11$</td><td>0.2</td><td>0.1</td><td></td><td></td><td></td><td></td><td></td></tr>
</table>

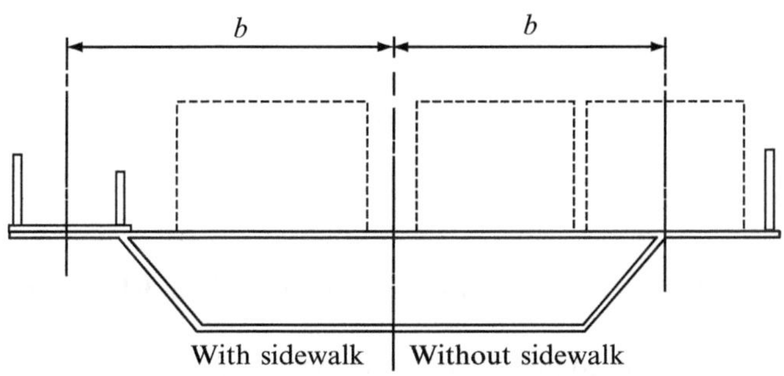

b: Distance between bridge center to center of sidewalk

b: Distance between bridge center to center of far-most lane

Fig. 2.4 Definition of b

(d) *Vortex-induced vibration in torsion*

$$U_{cv\theta} = 1.33 f_\theta \cdot B > U_d \tag{2.11}$$

And

$$\theta_c = \frac{E_\theta \cdot E_{t\theta}}{I_p \cdot \delta} \rho \cdot B^4 < \theta_a \tag{2.12}$$

where, θ_c = the maximum amplitude, and $\theta_a = 2.28/(b \cdot f_\theta)$ = the allowable amplitude. b is the distance between the bridge center and the center of the farthest lane or sidewalk as shown in Fig. 2.4.

Table 2.13 Coefficient $E_{t\theta}$

<table>
<tr><th rowspan="2">I_u / B/D</th><th colspan="10">Cross-section other than hexagonal shape</th><th rowspan="2">Hexa-gon</th></tr>
<tr><th>0.11</th><th>0.12</th><th>0.13</th><th>0.14</th><th>0.15</th><th>0.16</th><th>0.17</th><th>0.18</th><th>0.19</th><th>0.20</th></tr>
<tr><td>$2 < B/D \leqq 3$</td><td>0.7</td><td>0.6</td><td>0.5</td><td>0.4</td><td>0.3</td><td>0.2</td><td>0.1</td><td colspan="3"></td><td rowspan="9">1</td></tr>
<tr><td>$3 < B/D \leqq 4$</td><td>0.6</td><td>0.5</td><td>0.4</td><td>0..3</td><td>0.2</td><td>0.1</td><td colspan="4"></td></tr>
<tr><td>$4 < B/D \leqq 5$</td><td rowspan="2">0.5</td><td rowspan="2">0.4</td><td rowspan="2">0.3</td><td>0.2</td><td>0.1</td><td colspan="5"></td></tr>
<tr><td>$5 < B/D \leqq 6$</td><td rowspan="2">0.1</td><td colspan="6"></td></tr>
<tr><td>$6 < B/D \leqq 7$</td><td rowspan="3">0.4</td><td rowspan="2">0.3</td><td>0.2</td><td colspan="6"></td></tr>
<tr><td>$7 < B/D \leqq 8$</td><td rowspan="2">0.1</td><td colspan="7"></td></tr>
<tr><td>$8 < B/D \leqq 9$</td><td rowspan="2">0.2</td><td colspan="7"></td></tr>
<tr><td>$9 < B/D \leqq 10$</td><td rowspan="2">0.3</td><td colspan="8">0</td></tr>
<tr><td>$10 < B/D \leqq 11$</td><td>0.1</td><td colspan="8"></td></tr>
</table>

Also,

$$E_\theta = \frac{17.16\beta_{ds}}{(B/d)^3},$$

where β_{ds} is defined as (2.8) and $E_{t\theta}$ is defined in Table 2.13.

2.2.3.4 When There Are Concerns Regarding the Wind-Induced Vibration

After following the procedure described in Sect. 2.2.3.3, if there are concerns regarding the safety of the bridge in terms of wind-induced vibrations, the following steps are to be followed:

(a) Earlier wind tunnel test results on bridges with similar cross-sectional deck shapes are to be referred to, and a decision shall be made whether or not new wind tunnel tests are required for the present case.
(b) If further tests are deemed unnecessary, effort is to be made to modify the design parameters so that the results following the procedure described in Sect. 2.2.3.3 would be satisfactory. If this cannot be achieved, further wind tunnel tests are required.
(c) If the new wind tunnel test results indicate unsatisfactory performance of the deck cross-section, proper modification of it or the addition of aerodynamic appendices should be investigated in order to improve its aerodynamic performance. In case of phenomena, such as vortex-induced vibration, where the dynamic amplitude is limited in a range, once its probability of occurrence is properly addressed, the application of aerodynamic or structural remedy could be considered after the bridge is completed, unless any serious structural damages are anticipated immediately.

2.2.4 *Dynamic Wind Resistant Design of Very Flexible Members*

Some particular structural members are so flexible that a proper attention should be paid to their wind-induced dynamic behavior. They are

- Free-standing pylons for cable-stayed and suspension bridges
- Stay-cables for cable-stayed bridges
- Long and slender members of arch bridges and truss bridges
- Long and slender attachments such as light posts

2.2.5 *Dynamic Wind Resistant Design for Bridges During Erection*

Bridges at their erection stage particularly of the following cases would have significantly different conditions from those at their completion. Therefore, special wind-induced dynamic response considerations are required. Since the construction period is usually much shorter than the expected service period, the design wind speed for the construction stage can be reduced accordingly.

- When the pylons for cable-supported bridges are standing freely
- The cable-supported bridge decks before their closure, often at the mid-span
- When the cross-sectional shape of the deck differs from that of after completion

References

1. Japan Road Association (2002) Specifications for highway bridges (in Japanese). Maruzen, Tokyo
2. Japan Road Association (1991) Wind resistant design manual for highway bridges (in Japanese). Maruzen, Tokyo. Also revised version in 2007

Chapter 3
Wind Related Codes for the Honshu-Shikoku Project

As already mentioned in Chap. 1, the preparatory investigation, design and construction of the Honshu-Shikoku Bridges (HSB) had 40 years of history as shown in Table 3.1, and its construction finally came to the end by completing the Akashi-Kaikyo Bridge in 1998 and the Tatara Bridge as well as Kurushima-Kaikyo Bridges in 1999.

Though the stability of each of these bridges against wind was verified based on contemporary standards as shown in Table 3.1, there were significant research and development of new technology during this period in relation to the project. In order to summarize the outcome of this development through the whole period together with the accumulated experience, the results were integrated into the following two documents: *Wind Resistant Design Standard* (*2001*) and *Wind Tunnel Test Specification for Honshu-Shikoku Bridges* (*2001*). These two documents together with the *Wind Tunnel Test Specification for Akashi-Kaikyo Bridge* (*1990*) are introduced in the present chapter. The last document was actually used in preparation and design of the Akashi-Kaikyo Bridge.

3.1 Wind Resistant Design Standard for HSB (2001)

3.1.1 Outline

The final version of this *Standard* (2001) was established after all the design and construction for the project were actually completed, and in that sense, the document can be called as the final report on the wind-related works carried out for HSB. The chief objective of this final version is to prepare for the long bridges, such as suspension bridges with a span of 1,000 m or more and cable-stayed bridges with spans over 500 m, which may be projected in future. However, the *Standard* is, in principle, equally applicable to any other bridges of shorter spans with relatively insignificant modifications. Section 3.1 is to introduce this *Standard*.

Y. Fujino et al., *Wind Resistant Design of Bridges in Japan*,
DOI 10.1007/978-4-431-54046-5_3, © Springer 2012

Table 3.1 History of wind resistant design standards and construction of HSB

Year	Items	Construction and used design code (A)	(B)	(C)
1961	**Foundation of JSCE's advisory committee**			
1962				
1963	Foundation of Task Committee for wind resistant design			
1964	Guide Specification for Wind Resistant Design (1964)			
1965				
1966				
1967	Guide Specification for Wind Resistant Design (1967)			
1968				
1969				
1970	**Foundation of Honshu-Shikoku Bridge Authority**			
1971				
1972	Wind Resistant Design Standard (1972)			
1973				
1974				
1975	Wind Resistant Design Standard (1975)			
1976	**Wind Resistant Design Standard (1976)—(A)**			
1977				
1978				
1979				
1980				
1981				
1982				
1983				
1984				
1985				
1986				
1987				
1988				
1989				
1990	**Wind Resistant Design Spec. for Akashi-Kaikyo Bridge—(B)**			
1991	**Completion of large boundary layer wind tunnel facility**			
1992				
1993				
1994	**Wind Resistant Design Std for Onomichi-Imabari Route (1994)—(C)**			
1995				
1996				
1997				
1998				
1999				
2000				
2001	**Wind Resistant Design Standard (2001)**			

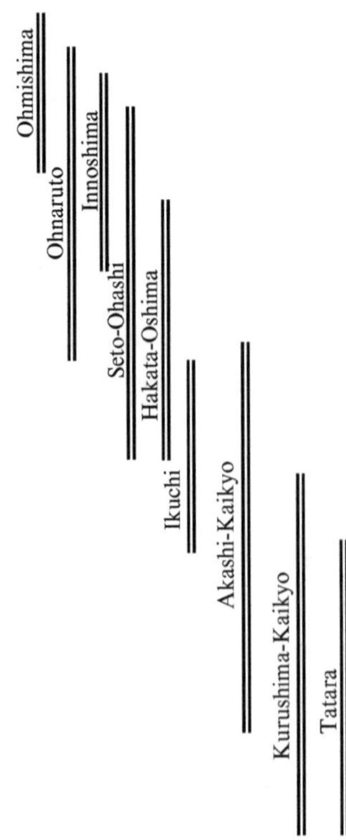

3.1.2 Scope of the Standard (2001)

This *standard* applies to the wind resistant design of the Akashi-Kaikyo Bridge, Tatara Bridge, and all of the Kurushima-Kaikyo Bridges. These bridges were actually designed based upon individually prepared design code for each bridge. The present *Standard* (2001) is a summary of all of these individual codes.

3.1.3 Definitions and Notations

3.1.3.1 Definition of Terminology

Basic wind speed	A 10-min mean wind speed at 10 m above the average sea level at the bridge site. It is given by the 150 year wind, which is the wind speed to be exceeded once during 100 years of expected service life with a probability of 50%
Design wind speed	A wind speed with which the applicable wind load is decided. Given by the basic wind speed multiplied by a factor depending upon the elevation of the structure
Critical wind speed	The lowest onset wind speed of a motion destructive to the structure
Reference wind speed	A benchmark speed against which the critical speed is examined
Design wind load	Wind induced drag force determined by the method defined in this section. Taken in two components: along and normal to the longitudinal bridge axis
Steady wind force	Time-averaged wind-induced force
Fluctuating force	Time-dependent wind-induced force caused by wind turbulence
Unsteady wind force	Motion-dependent wind-induced force. Taken in two components: in-phase and out-of-phase components associated with a simple harmonic body motion
Horizontal yaw angle	Directional angle of mean wind measured clockwise from the direction normal to the bridge
Inclination angle	Vertical angle of mean wind from the horizontal plane. Positive upward
Angle of attack	Vertical angle of mean wind from the lateral axis based on the bridge deck. Positive upward
Uniform flow	A wind flow in which any temporal velocity fluctuation is ignored
Turbulent flow	A wind flow in which the temporal and spatial velocity fluctuations are considered
Structural damping	Structure's capability to dissipate dynamic energy. Defined and expressed in terms of the logarithmic decrement
Basic sea level	The average water surface level of Tokyo Bay

3.1.3.2 Notations

Notations used in this document are defined as follows:

$\bar{U}, \bar{V}, \bar{W}$	Perpendicular components of the mean wind velocity in the main flow direction, the horizontal direction and the vertical direction, respectively
u, v, w	Components of the fluctuating wind velocity vector, two in horizontal and one in vertical directions
β	Azimuth angle
$\bar{U}_{10}$	Basic wind speed
$\bar{U}_z$	Design wind speed
P_d	Design wind load
ρ	Air density
C_d	Drag force coefficient
A_n	Projected area
μ_1	Modification factor for the design wind speed, considering the elevation of the structure
μ_2	Modification factor for the design wind load particularly for the cable systems and the suspended bridge deck, considering the effects of both wind and the structural characteristics
μ_3	Modification factor for the design wind load particularly for the pylons, considering the effects of both wind and the structural characteristics
μ_F	Modification factor for the reference wind speed of divergent vibrations considering the effects of turbulence

3.1.4 Wind Resistant Design Procedure

The wind resistant design is to be conducted along the following steps for each structural part (see also Fig. 3.1):

1. Approximate cross-sections of the bridge deck and towers are decided following the static design procedure.
2. Aerodynamically stable sections are chosen by the sectional model wind tunnel tests for the deck and by the aeroelastic model tests for the pylons, and the aerodynamic force coefficients are decided for the selected sections.
3. The static instability of the structure under wind action is examined.
4. Wind induced dynamic phenomena, divergent vibrations, gust induced response and vortex-shedding excitations, are examined by the aeroelastic full bridge model tests and/or the dynamic analyses.

Out of these steps, (3) and (4) can be skipped when the bridge span is medium or short and is designed with the structural configurations that are accepted to be aerodynamically stable.

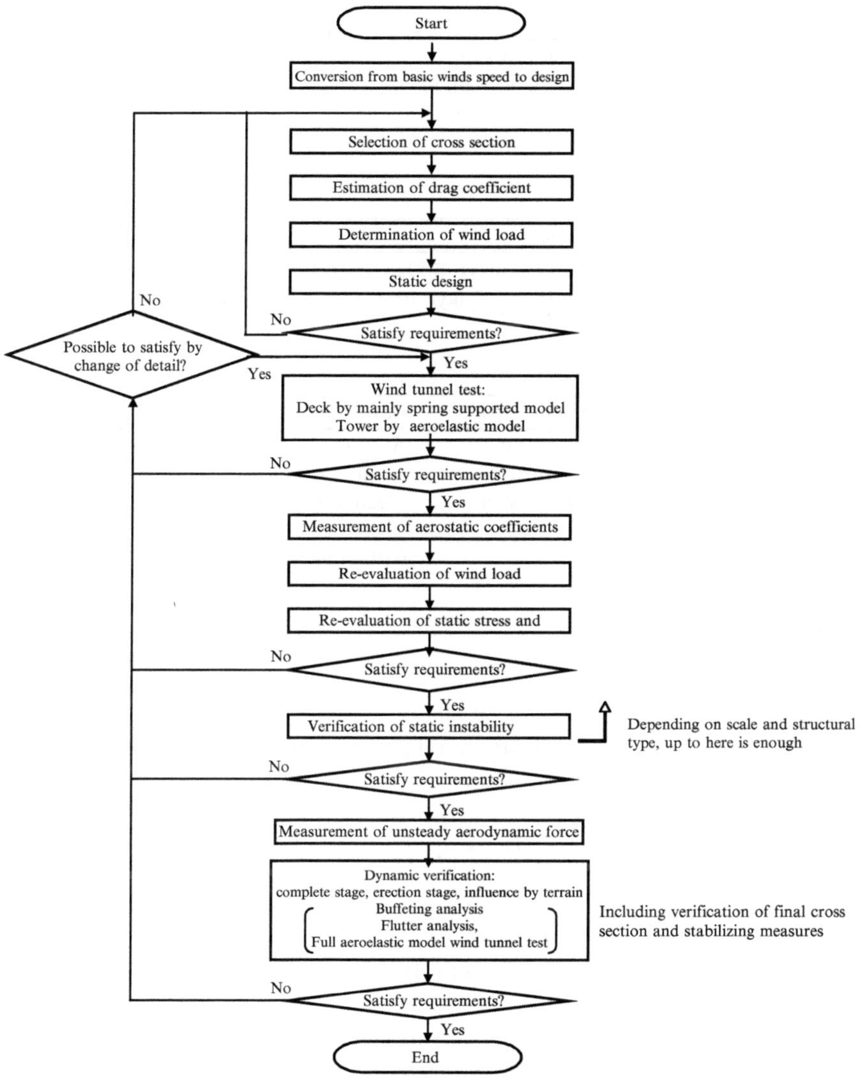

Fig. 3.1 Procedure of wind resistant design for the Akashi-Kaikyo Bridge, Tatara Bridge and Kurushima-Kaikyo Bridges

3.1.5 Basic Characteristics of Wind Considered in Design

3.1.5.1 Basic Wind Speed

The basic wind speed $\bar{U}_{10}$ for a few specific bridges are shown in Table 3.2. The background how these values were decided is provided in Sect. 4.3.

Table 3.2 Basic wind speed (at 10 m above the sea level)

Bridge	Basic wind speed (m/s)
Akashi-Kaikyo	46
Kurushima-Kaikyo	40
Tatara	37

Table 3.3 Exponent for the power law for the vertical profile of wind speed

Bridge	Exponent (α)
Akashi -Kaikyo Bridge	1/8
Kurushima-Kaikyo Bridges & Tatara Bridge	1/7

Table 3.4 Basic reference height of the structural parts

Structural part	Basic reference height, *z* (m)
Suspended structure	Average elevation of the stiffening or main girder at the main span
Cable system	Mean of the deck reference height and the tower top
Towers	65% of the tower height

3.1.5.2 Design Wind Speed

The design wind speed $\bar{U}_z$ for each structural part is defined by (3.1), considering its reference height.

$$\bar{U}_z = \mu_1 \cdot \bar{U}_{10} \tag{3.1}$$

where

$$\mu_1 = (z/10)^{\alpha} \tag{3.2}$$

in which, the power-law exponent α is given in Table 3.3 and the basic reference height for each structural part is shown in Table 3.4.

3.1.5.3 Characteristics of Wind Turbulence

Examination of Divergent Vibration

For the examination of the aerodynamic instabilities, turbulence intensities are assumed to be $I_u = 0.10$ in the along-wind direction and $I_w = 0.05$ for the vertical direction. These are decided based on the past measurement in very strong winds.

Examination of Vortex-Induced Vibration

For the examination of the vortex-induced vibration, the intensities of turbulence are assumed to be $I_u = 0.05$ in the along-wind direction and $I_w = 0.025$ in the vertical direction. These values are taken lower than the value for divergent vibration to be more conservative, since the vortex-induced vibration tends to be stronger in less turbulent wind.

Examination of Gust-Induced Response

The following turbulence characteristics are to be assumed for the examination of buffeting response:

- u-component: The power spectral density of the along-wind turbulence component, $S_u(f)$, is to be given by the Hino spectrum as follows:

$$\frac{f \cdot S_u(f)}{\sigma_u^2} = 0.475 \cdot \frac{f}{f'} \left\{ 1 + \left(\frac{f}{f'} \right)^2 \right\}^{-5/6} \tag{3.3}$$

  where f = frequency, $f' = 0.635z^{-0.75}$. z is measured in meters. The vertical profile of turbulence intensity is assumed to be

$$I_u(z) = I_u(10) \cdot \left(\frac{z}{10} \right)^{-\alpha} \tag{3.4}$$

  where α is the exponent given in Table 3.3.
- w-component: The power spectral density of the vertical component, $S_w(f)$, is to be given by the Bush–Panofsky spectrum as follows:

$$\frac{f \cdot S_w(f)}{\sigma_w^2} = \frac{2.11 \cdot f_r}{1 + 11.2 \cdot f_r^{5/3}} \tag{3.5}$$

  where $f_r = f \cdot z/\bar{U}_z$. The intensity of turbulence is assumed to be a half of the along-wind component.
- Spatial correlation of turbulence components is assumed to be expressed by an exponential function as follows:

$$coh_{ii}(f, r_j) = \exp\left(-k_{ij} \frac{f \cdot r_j}{\bar{U}_z} \right) \tag{3.6}$$

  where the decay factor k_{ij} can be taken as 8 for both along-wind and vertical directions. Refer to Sect. 4.2.4.2 for more description of the spatial correlation.

Table 3.5 Vertical inclination angle of wind

Air flow	Inclination
Uniform flow	– 3° to + 3°
Turbulent flow	0°

Influence of the Surrounding Topography

If the wind characteristics at the bridge site are considered to be strongly influenced by the surrounding topography, it is advisable to experimentally examine the local wind conditions by performing wind tunnel tests rather than following the above rules.

3.1.5.4 Vertical Inclination Angle and Horizontal Yaw Angle of Mean Wind

The vertical inclination of mean wind is to be considered in the range specified in Table 3.5. When a mean inclination angle of other than 0° is observed because of the surrounding topography, ±3° about the observed mean angle is to be considered. If a significant magnitude of static torsional deformation of the bridge deck is expected, the additional angle of attack relative to wind due to this deformation should be taken into consideration.

The mean wind direction is basically considered to be normal to the longitudinal bridge axis. However, if there is any reason to believe why other wind directions would give more serious effects on the bridge, the wind with these critical yaw angles should be also considered.

3.1.6 Static Design

3.1.6.1 Static Wind Load

For the static design, only the static drag is to be considered as the design wind load.

- Wind load on decks and cables:

$$P_D = \mu_2 \frac{\rho \bar{U}_z^2}{2} C_D A_n \tag{3.7}$$

- Wind load on pylons:

$$P_D = \mu_3 \frac{\rho \bar{U}_z^2}{2} C_D A_n \tag{3.8}$$

Table 3.6 Modification factors

Structural part		Suspended structure		Towers	
Direction		Transverse	Longitudinal	Transverse	Longitudinal
Akashi-Kaikyo Bridge					
Cables	μ_2	1.55	–	1.35	–
Suspenders & Stiffening truss	μ_2	1.55	1.25	1.35	1.25
Towers	μ_3	–	–	1.55	1.50 (Top fixed) 1.75 (Top freed)
Kurushima-Kaikyo Bridges					
Cables, Suspenders & Stiffening girder	μ_2	2.0	1.3	1.7	1.3
Towers	μ_3	–	–	1.4	1.55 (Top fixed) 1.7 (Top freed)
Tatara Bridge					
Cables & Girder	μ_2	1.9	1.35	1.65	1.35
Towers	μ_3	1.9	1.5	1.65	1.5 (Top fixed) 1.8 (Top freed)

where μ_2, μ_3 = the modification factors given in Table 3.6; ρ = the air density typically given by 1.18 kg/m^3; $\bar{U}_z$ = the design wind speed defined in Sect. 3.1.5.2; C_D = the drag force coefficient defined by Table 3.7; and A_n = the projected area defined by Table 3.8.

The drag force coefficient in this calculation must be verified by wind tunnel tests. The design calculation has to be repeated using the experimental value, if the assumed value for zero angle of attack was different from the experimental value by 5% or more. The way the modification coefficients μ_2, μ_3 are introduced is basically explained in Sect. 1.2.1.

3.1.6.2 Application of the Design Wind Load

The design wind load is generally considered to be a uniformly distributed horizontal load and is applied perpendicular to the longitudinal bridge axis. However, this loading pattern should be modified in a way so that the most disadvantaged effect is to be considered for each bridge member as follows:

Wind Load Perpendicular to the Bridge Axis

- Suspended deck: Loading on the windward side
- Towers: Loading on both windward and leeward shafts
- Suspension bridge cables: Uniformly along the cable axis
- Stay-cables: Equally divided to the suspended deck and the towers

Table 3.7 Drag force coefficient

Structural part		Transverse direction C_D	Transverse direction Note	Longitudinal direction C_D	Longitudinal direction Note
Suspended structure	Truss type	Based on result of wind tunnel test for similar shape		60% of C_D in transverse direction	When measured value on similar shape is available, use it
	Box girder type	Based on result of wind tunnel test for similar shape		30% of C_D in transverse direction	
Tower		1.8 or based on result of wind tunnel test for similar shape	Per one shaft of rectangular shaped tower	1.8 or based on result of wind tunnel test for similar shape	For rectangular shaped tower
Suspension cable		0.7		–	
Suspender rope		0.7		$0.7/\sqrt{2}$	For truss deck
				0.7	For box girder deck
Stay cable		0.7		0.7	

Table 3.8 The way to determine the projection area

Structural part	Transverse direction	Longitudinal direction
Suspended structure	Windward side of girder, curb and guard rails	Same as transverse
Towers	Windward and leeward shafts	Both shafts and web members
Suspension cables	Windward and leeward cables	Neglect
Hanger ropes	Windward and leeward suspender ropes	Same as transverse
Stay cables	Windward and leeward cables	Same as transverse

Table 3.9 Increase of allowable stress for the load combination

	Increase factor of allowable stress	
Combination of loads	Tower	Suspended structure
D+W+T+SD+E	1.40	1.50

where, D: Dead load, W: Wind load, T: Influence of temperature change (when it is combined with the design wind load, the temperature shall be 35°C), SD: Influence of movement of supporting points, and E: Influence of error in fabrication and erection of superstructure.

Wind Load Along the Bridge Axis

- Suspended deck: Uniformly in longitudinal direction
- Towers: Total wind load, which is decided by entire projection area of the shafts and web members, is to be uniformly distributed
- Suspension bridge cables: Wind load on the cable itself is not considered, but a half of the load on hangers is to be taken into account
- Hanger ropes: Equally divided to the deck and the cables
- Stay-cables: Equally divided to the deck and the towers

Influence of the Horizontal Yaw Angle

Simultaneous loading both in perpendicular and longitudinal directions to the bridge axis is to be considered for design of shoes, expansion joints and stay-cables etc., where the longitudinal displacement would be of a concern.

3.1.6.3 Combination of Loads and Increase of Allowable Stress

An increase of the allowable stress is applied when the wind load is combined with the other loads as specified in Table 3.9.

Table 3.10 Structural damping (logarithmic decrement)

Vibration mode and structural part		Logarithmic decrement: Bending vibration	Logarithmic decrement: Torsional vibration
Deck-vibration dominant mode	Stiffening truss	0.03	0.02 (can be 0.03 for span of 500–600 m or less)
	Closed-box girder	0.02	0.02
Tower-vibration dominant mode	Tower-cable-girder system	0.02	0.02
	Free-standing tower	0.01	0.01
Stay-cable-vibration dominant mode		0.003	–
Suspender-rope-vibration dominant mode		0.003	–

3.1.7 Examination of Dynamic Behavior

Once the bridge is designed against static wind load, a possibility of wind-induced dynamic behavior must be also examined. The basic principle of examination against wind-induced dynamic behavior is described in this section. If two or more bridges are located in parallel, a possible aerodynamic interference between them has to be also examined, though this issue is not specifically mentioned in this section.

3.1.7.1 Structural Damping

Structural damping of a bridge in planning is not yet known when analyzing its dynamic behavior, though its magnitude may have a significant influence on the conclusion of the analysis. Under these circumstances, it is advisable to use the empirical damping values suggested in Table 3.10. If any artificial dampers are attached to the structure, their influence is to be evaluated separately.

3.1.7.2 Examination of Divergent Vibration

The critical wind speed of divergent vibration is to satisfy the following condition:

$$U_F \geq U_r = 1.2 \cdot \mu_F \cdot \bar{U}_z \tag{3.9}$$

where $\mu_F =$ the modification factor considering the velocity fluctuation, defined in Table 3.11; $\bar{U}_z =$ the design wind speed.

Table 3.11 Modification factor for divergent type vibration of deck by fluctuating wind speed

Bridge	μ_F
Akashi-Kaikyo	1.08
Tatara	1.10
Kurushima-Kaikyo	1.10

Table 3.12 Basic wind speed during erection work

Bridge	Basic wind speed during erection work (m/s)
Akashi-Kaikyo	37
Kurushima-Kaikyo	31
Tatara	29

3.1.7.3 Examination of Gust Induced Response

The prediction of gust induced response is to be carried out both by the buffeting analysis and by physical model tests using simulated atmospheric boundary layer flow. The bridge has to be redesigned if the predicted stresses due to this dynamic response exceed the allowable stress by 5% or more.

3.1.7.4 Examination of Vortex-Induced Vibration

The suspended bridge deck and other bridge members are to be carefully examined so that any serious extent of vortex-induced oscillations would be avoided. The vortex excitation of towers in particular, together with its suppression, is discussed in Sect. 9.2.

3.1.7.5 Examination of Cable Vibration

The cables and hanger ropes are to be carefully examined so that any serious extent of their vibration would be avoided as much as possible. The suppression of cable vibration is discussed in Sect. 10.4.

3.1.8 Bridge Safety During Erection

3.1.8.1 Basic Wind Speed for Erection Design

The basic wind speed for the examination of bridge safety during its erection can be reduced from that is defined in Sect. 3.1.5. Table 3.12 lists the reduced basic wind speeds for three cases. These wind speeds were decided assuming the probability of not exceeding these speeds to be 0.80 for the given erection period, which was 5 years for Akashi and 4 years for Kurushima and Tatara.

3.1.8.2 Items to Be Studied

Bridges under construction conditions are generally more vulnerable to wind action compared to the conditions after their completion. In particular, the stability of free-standing towers before erection of cables, and suspended decks during their erection must be carefully examined and vibration control should be considered if there is any possibility of having instability.

3.2 Wind Tunnel Test Specifications for Honshu-Shikoku Bridges

A full-dress investigation on the Honshu-Shikoku Bridge project started in around 1960. Several task committees were formed under the auspices of the Japan Society of Civil Engineers to investigate technical challenges met by the long-span bridges, particularly on the fast tidal current, submarine foundation, strong wind, earthquakes, installation of high-speed trains, etc. A subcommittee for wind resistant design of the Honshu-Shikoku Bridges started in 1963 and submitted a report in 1967 including *Guide Specification for Wind Resistant Design of Honshu-Shikoku Bridges (1967)*.

When the wind resistant design is considered, the most contemporary idea in those days was that the wind tunnel tests would be essential for determining the static force coefficients of the bridge deck section and also for the examination of safety against wind-induced vibration, where the accuracy of the wind tunnel tests had to be guaranteed. By that time, many wind tunnel tests had been carried out in Japan and engineers had good experience and understanding of the subject matter. They understood that:

1. Wind tunnel tests of the original Tacoma Narrows Bridge were able to reproduce the wind-induced vibration observed in reality
2. Wind tunnel tests of the Golden Gate Bridge was able to reproduce the wind-induced vibration observed in full-scale at the wind speed of 31 m/s
3. Wind tunnel tests repeated at different institutions showed a similar tendency of the test results
4. Suspension bridges which were designed based on wind tunnel tests were able to avoid accidents or damage due to adverse wind effects
5. The effectiveness and validity of wind tunnel tests were confirmed based on these points. *Guide Specification for Wind Resistant Design of Honshu-Shikoku Bridges (1967)* then specified general requirements for wind tunnel tests

Afterward, the subcommittee reevaluated the procedures of wind tunnel tests and established a new *Standard of Wind Tunnel Test for Honshu-Shikoku Bridges* in 1971, followed by its revision in 1973, including the wind tunnel test procedure for towers. These standards considered only the truss girder stiffened suspension bridges. After the scope was expanded to a possibility of having box girders, cable-stayed bridges and other type bridges, *Wind Tunnel Test Specification for Honshu-Shikoku Bridges (1980)* was finally formed. It covers the procedures for static force measurements, spring-supported section model tests and tower tests, and specifies model fabrication, similitude requirements, various measuring methods and also a standard way of test result presentation.

Technical challenges were posed by the Innoshima Bridge, Ohnaruto Bridge and Seto-Ohashi until early 1980s, then came the world's longest suspension bridge, the Akashi-Kaikyo Bridge. For the Akashi-Kaikyo Bridge, there were concerns regarding a possibility of wind-induced vibration of the towers even after the completion, which had an unprecedented height of 300 m. *Wind Tunnel Test Specification for Akashi-Kaikyo Bridge (1990)* added the test procedures of the towers in service conditions and also the taut strip model tests.

The Honshu-Shikoku Bridge Authority decided to carry out a large-size aeroelastic full bridge model tests for the Akashi-Kaikyo Bridge (1998), the Tatara Bridge (1999) and the Kurushima-Kaikyo Bridges (1999), because consideration of three dimensionality (spatial non-uniformity) of airflow and structures in complex terrains, particularly for the case of the Tatara and Kurushima-Kaikyo Bridges, was deemed important. Before the full bridge model tests, there was no written provision regarding the wind tunnel tests of 3D aeroelastic model. Therefore, the tests were conducted based on the discussions in committees and working teams.

In addition, flutter analyses were conducted in order to predict the flutter onset wind speed of the bridge with final design. This is because the full aeroelastic model used in the wind tunnel test generally had slightly different configuration from the final design and re-evaluation of the test results was necessary based on the analysis. Unsteady aerodynamic force coefficients were measured to obtain necessary information for flutter analysis.

In 2001 after the construction of all the Honshu-Shikoku Bridges was completed, all wind tunnel test procedures learned and developed through the Honshu-Shikoku Bridge project were assembled together to be *Wind Tunnel Test Specification for Honshu-Shikoku Bridges (2001)*. Full aeroelasitic bridge model test procedure and measurement of unsteady aerodynamic force coefficients were added in this version.

Some contents extracted from these specifications are given in Appendix 1.

Chapter 4
Design Wind Speed

4.1 Introduction

In this chapter, the wind characteristics described in the design codes are explained. First, the general characteristics of natural wind are explained, and various parameters used in the two representative Japanese bridge design standards are defined. Finally, the effects of local topography on the design wind characteristics are introduced as an example of recent development in wind resistant design methods.

4.2 Wind Characteristics

Wind is characterized by various different parameters. They include the following items: (1) wind speed, such as the mean wind speed and maximum instantaneous wind speed; (2) wind direction such as the azimuth direction and the angle of vertical inclination; (3) time-dependent fluctuation of the velocity vector expressed in terms of turbulence intensities and scales of turbulence, or the shape of the velocity spectra; and (4) the spatial variation characteristics such as wind speed profiles and spatial correlations. This section details each of these items, and some examples of the measured field data are presented.

4.2.1 Mean Wind Speed and Instantaneous Wind Speed

Wind speed is often defined in two ways: the "mean wind speed", which is an average of the instantaneous wind speed over a certain averaging period, and the "maximum instantaneous wind speed", which indicates the recorded maximum value within a given period.

Y. Fujino et al., *Wind Resistant Design of Bridges in Japan*,
DOI 10.1007/978-4-431-54046-5_4,

$$\text{Mean wind speed} = \frac{1}{T}\int_{t=0}^{T} u(t)\,dt \qquad (4.1)$$

$$\text{Maximum instantaneous wind speed} = \max[u(t)],\ 0 \le t \le T \qquad (4.2)$$

where $u(t)$: wind speed at time t, T: averaging time.

In Japan, a period of 10 min has been adopted as the averaging time. The ratio of the maximum instantaneous wind speed to the mean wind speed, shown in the following formula, is defined as the "gust factor". The gust factor is a factor to indicate the extent of the wind speed fluctuation in a given period.

$$\text{Gust factor} = \text{Max instantaneous wind speed}/\text{Mean wind speed} \qquad (4.3)$$

4.2.2 Wind Direction and Angle of Inclination

By setting a standard horizontal axis x' as in Fig. 4.1, the wind direction β is defined as the angle between the wind velocity vector in the $x'y'$-plane and the x'-axis, and the angle of inclination θ is defined as the angle between the wind velocity vector and the $x'y'$-plane, respectively.

By defining U, V, and W as the components of mean wind velocity vector in x', y', and z directions, respectively, as shown in Fig. 4.1, the direction of mean wind β and the angle of inclination θ can be calculated as follows:

Wind direction:

$$\beta = \arctan\left(\frac{V}{U}\right) \qquad (4.4)$$

Angle of inclination:

$$\theta = \arctan\left(\frac{W}{\sqrt{U^2 + V^2}}\right) \qquad (4.5)$$

Other expressions are also possible as follows for β and θ, where the vector averaging is used. β' and θ' are called the mean direction of wind and mean angle of inclination, respectively. But β and θ obtained by scalar averaging are usually used as the mean values because they are easier for data processing.

Wind direction (vector averaging):

$$\beta' = \int_{t=o}^{T} \arctan\left(\frac{v(t)}{u(t)}\right) dt/T \qquad (4.6)$$

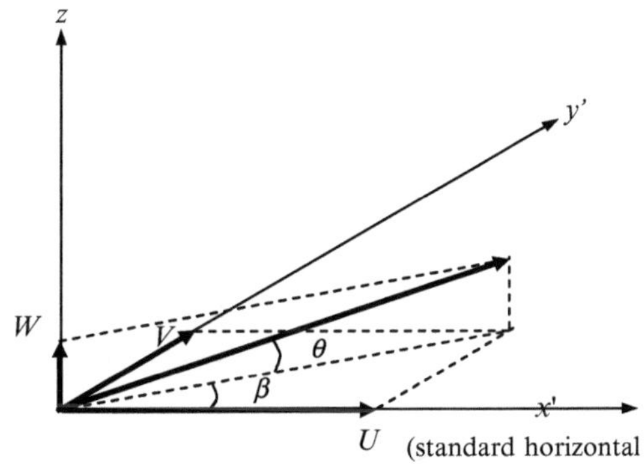

Fig. 4.1 Definition of wind direction and the angle of inclination

Angle of inclination (vector averaging):

$$\theta' = \int_{t=0}^{T} \arctan\left(\frac{w(t)}{\sqrt{u(t)^2 + v(t)^2}}\right) dt/T \tag{4.7}$$

Incidentally, the angle of attack used for the section model wind tunnel tests is the 'relative angle of wind against the deck'. It is determined by considering the angle of inclination (defined above), pitching deformation of the deck, angle of inclination due to vertical component of wind velocity fluctuation, and possible experimental error. Related discussions are also given in Sect. 6.4.2.

4.2.3 *Intensities, Scales and Spectra of Turbulence*

4.2.3.1 Turbulence Intensities

As shown in Fig. 4.1, a wind velocity vector is resolved into three directional components, and each directional component consists of its mean (U, V, W) and fluctuating part (u, v, w).

By defining the standard horizontal axis x' in such that $V = 0$, and also assuming that the vertical component of mean wind is negligibly small, the turbulence intensity in each direction is defined as follows:

$$I_u = \frac{\sigma_u}{U},\ I_v = \frac{\sigma_v}{U},\ I_w = \frac{\sigma_w}{U} \tag{4.8}$$

where σ_u,σ_v, and σ_w are the standard deviation of u, v and w, respectively.

Examples of wind data measured in the field are shown in the following.

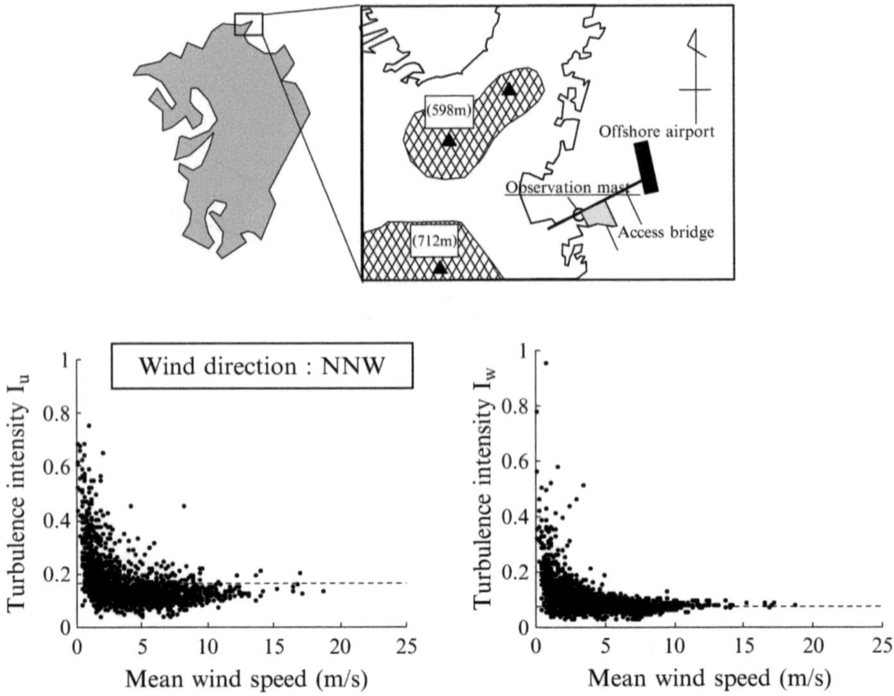

Fig. 4.2 Measured turbulence intensities (Shin-Kitakyushu Airport) [1]

Shin-Kitakyushu Airport Access Bridge

An example of field measured turbulence intensities is shown in Fig. 4.2. The data were obtained at 30 m above ground near the seacoast by using an ultrasonic anemometer [1].

From Fig. 4.2, it is found that the turbulence intensities are scattered, especially when the mean speed is low but tend to converge to a certain value at higher wind. Conversion of *w*-component seems to take place at lower wind speed compared to that of *u*-component.

For the wind resistant design, the definition of turbulence intensity would be needed in fairly high wind speed. Therefore, if the data in high wind speed range are not available and only the data from low wind are taken into statistics, consideration is required because it often results in overestimating turbulence intensity. On the other hand, attention also has to be paid, because very small turbulence intensity is sometimes observed even with relatively low wind speed in stable condition under thermal stratification.

The Ohnaruto Bridge and the Akashi-Kaikyo Bridge

Wind speed measurement was carried out at 12 m above the road surface near the mid-point of the center span of each bridge, using five ultrasonic anemometers for

Fig. 4.3 Site of the Ohnaruto Bridge and the Akashi-Kaikyo Bridge [2, 3]

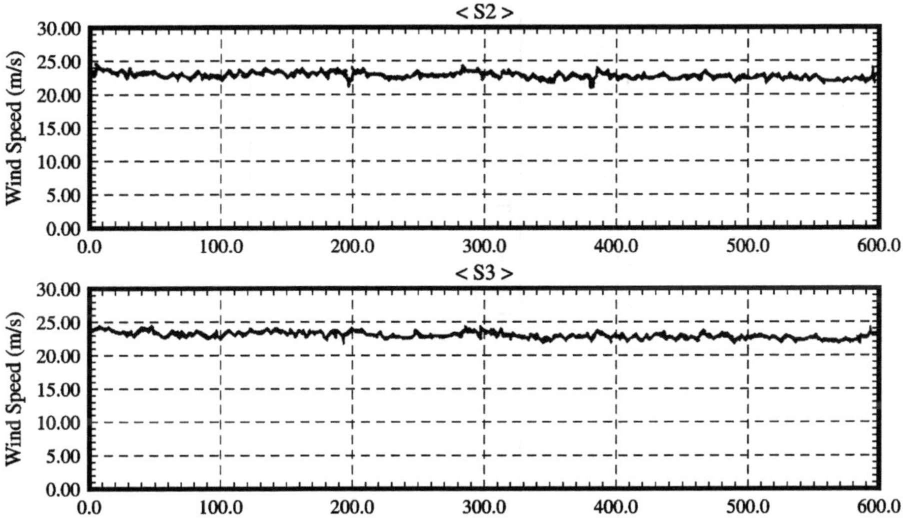

Fig. 4.4 Time history of wind speed measured at the Ohnaruto Bridge site ('96/03/17 02:59–03:09, wind direction S) [2, 3]

Ohnaruto and five windmill-type anemometers for the Akashi-Kaikyo Bridge. The ultrasonic anemometer S3 locates at the center span and S2 locates 10.6 m away from S3 on the Ohnaruto Bridge, at the height of 74 m above the sea surface. Each bridge site is shown in Fig. 4.3 and measured results are shown in Fig. 4.4 and Table 4.1 [2, 3].

By focusing upon the seasonal wind data measured at the Ohnaruto Bridge on March 17, 1996, the longitudinal turbulence intensity is found extremely low, only at the order of 2%. This was a precious measurement which showed that almost uniform flow can exist in an open sea environment. For the full bridge model tests to examine the resistance against vortex-induced vibration, 5% turbulence intensity for the approaching flow at the deck height is to be assumed as described

Table 4.1 Measured 10-min mean wind speed and turbulence intensity (Ohnaruto Bridge and Akashi-Kaikyo Bridge) [2, 3]

	At Ohnaruto Bridge				At Akashi-Kaikyo Bridge			
	Occasions	Time	10 min mean	Wind direction	Occasions	Time	10 min mean	Wind direction
U_{mean} (m/s)	Seasonalwind on	3:29–3:39	22.8	Southern wind almost	Typhoon 9807 on	14:13–14:23	32.1	Changing as movement
U_{max} (m/s)	Mar.17/1996		24.5	normal to the	Sep. 22/1998		38.9	of the typhoon almost
I_u (%)			2.08	bridge axis			7.02	45° from the normal
U_{mean} (m/s)		3:39–3:49	22.6			14:23–14:33	33.2	of the bridge axis
U_{max} (m/s)			24.3				39.1	
I_u (%)			1.98				6.26	
U_{mean} (m/s)		3:49–3:59	22.1			14:43–14:53	28.3	
U_{max} (m/s)			24.5				41.8	
I_u (%)			2.30				6.38	

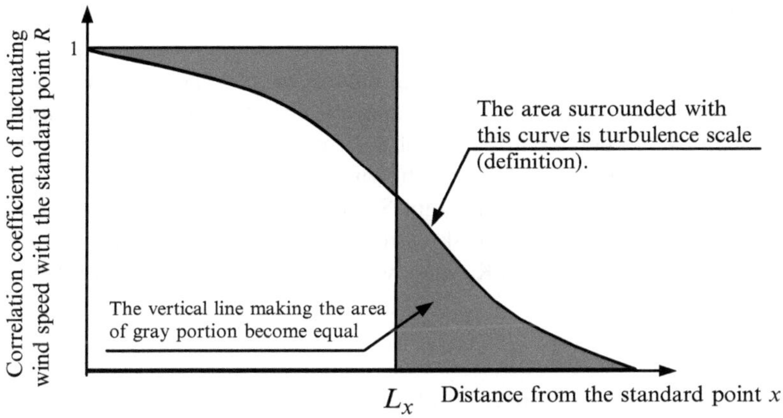

Fig. 4.5 Concept of the scale of turbulence

in Sect. 3.1.5.3. The value of 5% was adopted so that the test results will be done with conservative conditions. Note that the standard turbulence intensity for other tests is assumed to be 10%. However, the observation above indicates that the wind tunnel tests using 5% turbulence may not be conservative enough. In fact, it may be more appropriate to examine the stability against vortex-induced vibration in less turbulent flow if the very small turbulence intensity like above is to be considered.

4.2.3.2 Scales of Turbulence

The scale of turbulence of i-component of the velocity vector (where $i = u, v$ or w) in j-direction (where $j = x', y'$, and z), L_j^i, is defined as follows:

$$L_j^i = \int_0^\infty R_i(r_j)\, dr_j \tag{4.9}$$

where $R_i(r_j)$ is the correlation coefficient of fluctuating i-component velocity between two points, 'j' and '$j + r_j$', which is defined as follows:

$$R_i(r_j) = \frac{1}{\sigma_i(j) \cdot \sigma_i(j + r_j)} \sqrt{\frac{1}{T} \int_0^T i(j) i(j + r_j) dt} \tag{4.10}$$

The scale of turbulence indicates the spatial extent of fluctuating wind speed, which can be easily understood by considering an 'equivalent distance for which the wind fluctuation is fully correlated' (see Fig. 4.5).

4.2.3.3 Shape of Velocity Spectra

Fluctuation of natural wind is random in nature, so its standard deviation consists of various periodic components. Power spectral density function of the fluctuating wind velocity represents contribution of each periodic component to the standard deviation, and various models in each wind direction (u, v and w) have been proposed. In Wind Resistant Design Standards for Honshu-Shikoku Bridges, Hino's model and Busch & Panofsky's model are used for u- and w-components, respectively. The model by von Kármán, which is widely known, is also listed in the following:

Hino's Model (One-Sided Spectrum) [4, 5]

$$\frac{f \cdot S_u}{\sigma_u^2} = 0.4751 \cdot \frac{f}{f'} \cdot \left\{ 1 + \left(\frac{f}{f'} \right)^2 \right\}^{-\frac{5}{6}} \tag{4.11}$$

$$f' = 1.7181 \cdot 10^{-2} \cdot \frac{\alpha \cdot K_r \cdot U_{10}}{I_u^3} \cdot \left(\frac{z}{10} \right)^{(2m-3)\alpha - 1} \tag{4.12}$$

where K_r = the surface friction factor, m = a correction coefficient, α = the power-law exponent for the mean wind speed profile in vertical direction; and U_{10} = 10-min mean wind speed at 10 m above ground

von Kármán's Model (One-Sided Spectrum) [6, 7]

$$\frac{f \cdot S_u}{U_z} = \frac{4n_u \cdot I_u^2}{\left(1 + 70.8 \cdot n_u^2\right)^{5/6}} \tag{4.13}$$

$$\frac{f \cdot S_v}{U_z} = \frac{4n_v \cdot \left(1 + 755.2n_v^2\right) \cdot I_v^2}{\left(1 + 283.2 \cdot n_v^2\right)^{11/6}} \tag{4.14}$$

$$\frac{f \cdot S_w}{U_z} = \frac{4n_w \cdot \left(1 + 755.2n_w^2\right) \cdot I_w^2}{\left(1 + 283.2 \cdot n_w^2\right)^{11/6}} \tag{4.15}$$

where S_i = the power spectral density function of the i-component velocity; I_i = turbulence intensity in i-direction; U_z = the mean wind speed at height z above ground; $n_i = \frac{L_{x'}^i \cdot f}{U_z}$; $L_{x'}^i$ = the scale of turbulence of i-component velocity in x'-direction; and f = the frequency of wind velocity fluctuation.

Busch & Panofsky's Model (One-Sided Spectrum) [8]

$$\frac{f \cdot S_w}{\sigma_w^2} = \frac{0.632 \cdot \left(\frac{f_r}{f_{r\,\max}}\right)}{1 + 1.5 \cdot \left(\frac{f_r}{f_{r\,\max}}\right)^{\frac{5}{3}}} \tag{4.16}$$

in which $f_r = \frac{f\,z}{U_z}$; $f_{r\,\max} = f_r$ where the spectrum takes its maximum peak. The peak valuc of 0.3 shows good agreement with the field results.

4.2.4 *Wind Speed Profile and Spatial Correlations*

Spatial variation characteristics are divided into wind speed profiles, which indicate the distribution of mean wind speed, and spatial correlation that indicates the correlation of fluctuating wind speeds at two different locations.

4.2.4.1 Mean Wind Speed Profile

Wind speed profile can be defined both in horizontal and vertical directions, but horizontal profile is considered negligible unless the local topographic effect is predominant. Therefore, the wind speed profile only in vertical direction is usually used for wind resistant design.

The distribution of mean wind speed in free atmosphere above the gradient height is considered to be nearly homogeneous since the flow is independent of the ground surface conditions. However, in the boundary layer, the flow is retarded because of the frictional force. Hence, the lower the elevation is, the less the mean speed is expected to be. The vertical profile of mean wind speed inside the neutral boundary layer is often modeled either by the logarithmic law or the power law. The use of the power law is more common in Japan.

$$\frac{U_z}{U_G} = \left(\frac{z}{z_G}\right)^{\alpha} \quad (z \leq z_G) \tag{4.17}$$

$$\frac{U_z}{U_G} = 1 \quad (z > z_G) \tag{4.18}$$

where U_z = the mean wind speed at height z; U_G = the mean wind speed at the gradient height; z_G = the gradient height (or the boundary layer thickness); α = the power law exponent.

Table 4.2 Measured power law exponents of strong winds (Nakagawa River) [9]

	Wind direction	NNW	N	NNE	NE	ENE	E	ESE	SE	SSE	S	SSW	NW	WNW
Typhoons	Number of data	5	1	14	2	1	6	2	20	16	13	1	9	8
	n(=1/p)	5.4	6.4	4.5	4.3	3.9	6.4	5.1	6.7	5.2	4.8	5.2	5.6	7.3
Others	Number of data	107	32	1				8	17	43	38		151	7
	n (=1/p)	10.1	14.1	13.9				5.9	5.0	5.8	4.5		9.0	8.6
		← From the sea →								← From the land (Undulated) →			From the land (Smooth)	

When z is less than the representative height of the surface roughness, however, a different profile is usually assumed.

Various values are adopted for the power law exponents by referring to the field measurement data and other established standards. Examples of field measurement data are shown in the following:

The Coast of Nakagawa River [9]

(Observation height was up to 150 m above the ground level)

Calculated power law exponents at the Nakagawa River coast are shown in Table 4.2. In the cases of the wind from seaside, the power law exponent is from 1/3 to 1/7 for typhoon winds, and from 1/5 to 1/14 for other strong winds.

Tarumi Steel Tower [10]

(Observation height was up to 80 m above the ground level)

Vertical wind speed profile for each wind direction is shown in Fig. 4.6, which is calculated based on observed strong wind data. From these data, the value of 1/8 to 1/7 can be adopted for the power law exponent, representing the vertical wind speed profile at the center of the strait by the data of eastern wind direction along the coast. The continuous observation for 20 years at Tarumi steel tower (Fig. 4.7) formed the basis of the design wind speed for Honshu-Shikoku Bridges.

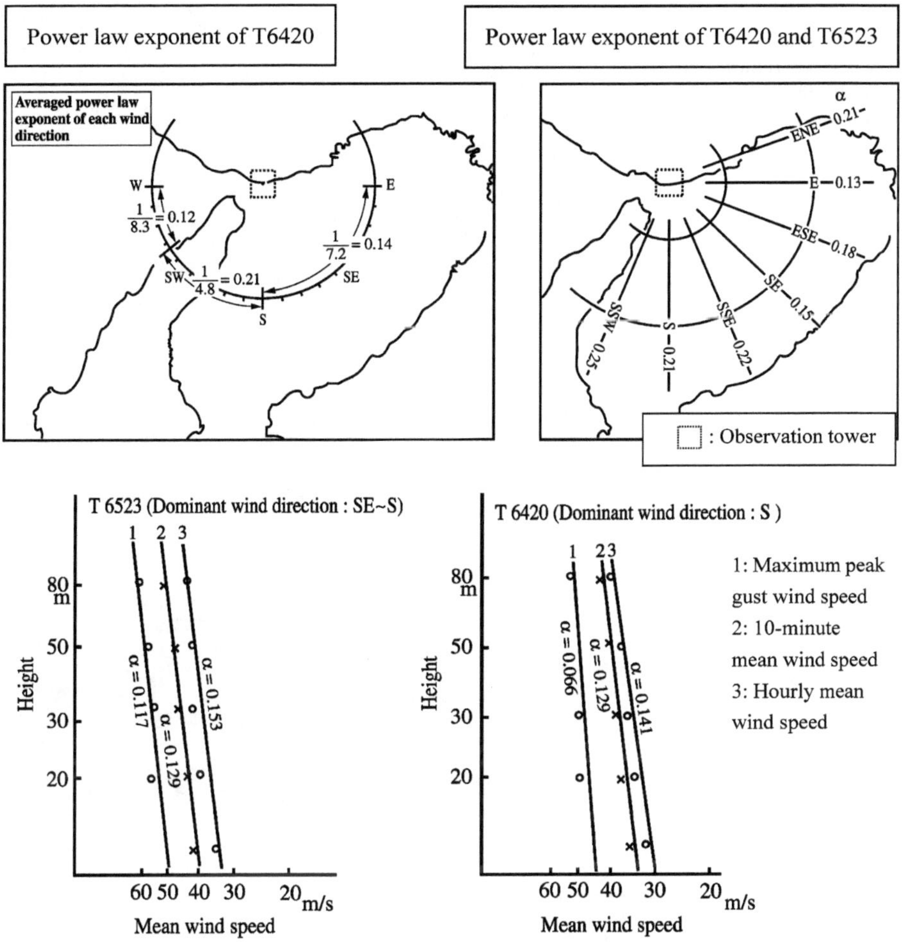

Fig. 4.6 Measured vertical wind speed profile of the typhoons (Tarumi) [10]

Station "C" at the Kansai International Airport [11]

(Observation height was 10–100 m above the sea level)

Calculated power law exponents are shown in Fig. 4.8. There is a tendency that the power law exponents in case of wind from the landside are different from winds from the seaside. The former gave 1/3 to 1/4, and the latter showed 1/4 to 1/5. The measured data were used to determine the design wind speed for the Kansai International Airport Access Bridge and other bridges near the site of this bridge.

Fig. 4.7 Tarumi steel tower for the wind speed observation (courtesy of the Honshu-Shikoku Bridge Authority)

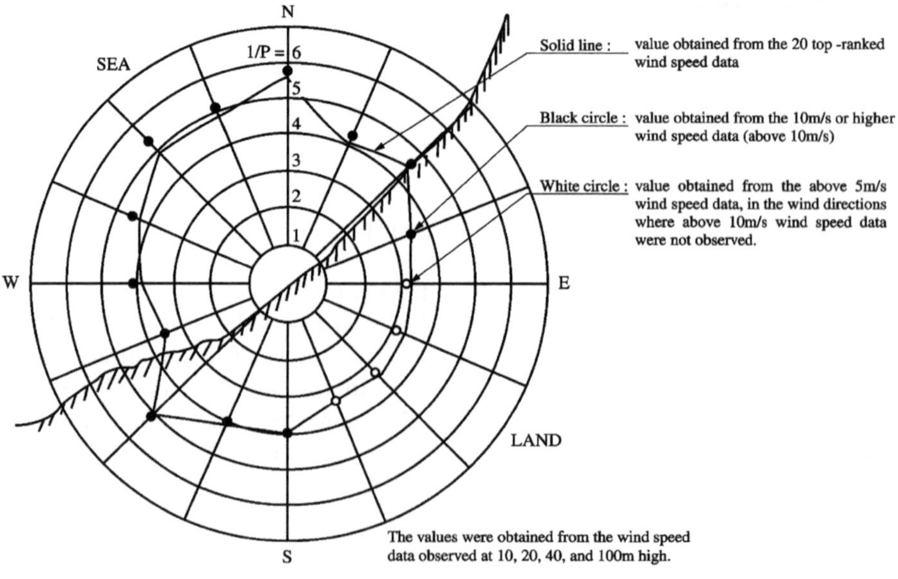

Fig. 4.8 Measured power law exponents (C station of the Kansai International Airport) [11]

Table 4.3 Measured power law exponents at the Onaruto Bridge site [12]

Wind direction	Power law exponent	Remarks
North wind	1/15–1/4	Three cases, $U_{67} = 14.2$–21.0 m/s
South wind	1/36–1/3.5	Fourteen cases, $U_{67} = 14.1$–21.0 m/s
	Average 1/9.5	Average of the middle eight cases from the data (1/15–1/6.3)

The Ohnaruto Bridge [12]

(Observation height was from the mid-height to the top of the tower on the Shikoku side (144 m))

Calculated power law exponents are shown in Table 4.3. The mean exponent was found to be approximately 1/10. This measurement was conducted after the completion of the Ohnaruto Bridge to confirm the validity of wind speed characteristics adopted in the design standards. Although the variation of the observed data was large, the adopted values in the standards are considered to be appropriate.

4.2.4.2 Spatial Correlations

The instantaneous wind speeds measured simultaneously at two different points are generally different from each other, even if their statistical properties are the same. The root-coherence function of two velocity components is defined as the correlation coefficient of these two data as a function of frequency as follows:

$$coh_{ii}(f, r_j) = \frac{|S_{ii}(j, j + r_j, f)|}{\sqrt{S_i(j, f) \cdot S_i(j + r_j, f)}} \tag{4.19}$$

where $S_{ii}(j, j + r_j, f)$ = the cross-spectral density function of i-component wind velocity at two points located, separated by r_j in j-direction.

Although a few models are proposed for the function, the Davenport model by the use of an exponential function, is usually applied in Japan.

Davenport Model [13]

$$coh_{ii}(f, r_j) = \exp\left(-k_{ij}\frac{f \cdot r_j}{U_z}\right) \tag{4.20}$$

where k_{ij} = a decay factor. As the value for k_{ij}, 7 (*Wind Resistant Design Standard for Honshu-Shikoku Bridges (1976)*) or 8 (*Wind Resistant Design Specification for Akashi-Kaikyo Bridge (1990)*) are adopted. Moreover, the following model has been applied to buffeting analysis in order to reproduce small correlation at lower frequency range when two points are located in the distance.

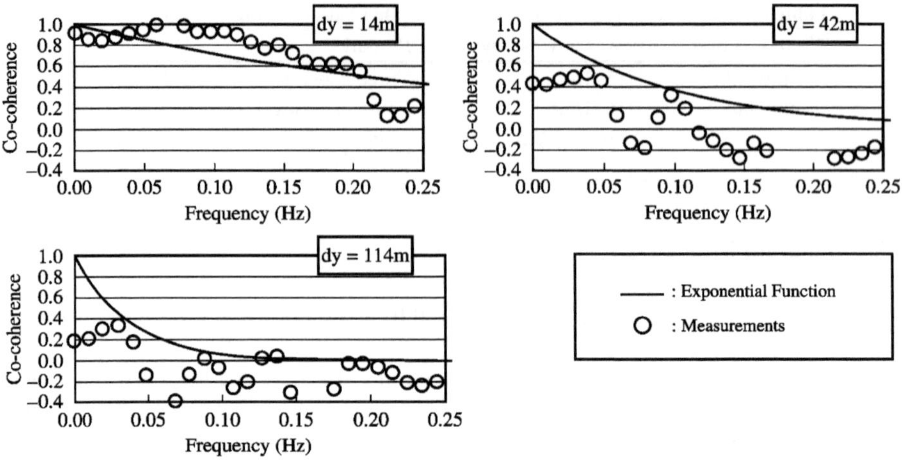

Fig. 4.9 Measured spatial correlation (the Akashi-Kaikyo Bridge) [3]

von Kármán's Model [6]

u-component:

$$coh_{ii}(f, r_j) = 0.994\left\{\eta^{5/6}K_{5/6}(\eta) - \frac{\eta^{5/3}}{2}\left(\eta^{1/6}K_{1/6}(\eta)\right)\right\} \tag{4.21}$$

$$\eta = \frac{r_j}{L_x^u}\cdot 0.747\sqrt{1 + 70.8\left(\frac{f L_x^u}{U_z}\right)^2} \tag{4.22}$$

w-component

$$coh_{ii}(f, r_j) = 0.994\left\{\eta^{5/6}K_{5/6}(\eta) - \frac{\eta^{11/6}K_{1/6}(\eta)}{1 + 188.7\cdot\left(\frac{f L_x^w}{U_z}\right)}K_{1/6}(\eta)\right\} \tag{4.23}$$

$$\eta = \frac{r_j}{L_x^w}\cdot 0.747\sqrt{1 + 70.8\left(\frac{f L_x^w}{U_z}\right)^2} \tag{4.24}$$

where K_ν = the second kind modified Bessel function of order ν.

Examples of the field measurement data at the Akashi-Kaikyo Bridge [3] are shown in the following:

The field measurement was carried out at the bridge site after completion, and the results shown in Fig. 4.9 were obtained by analyzing some data of strong winds. It was confirmed that the correlation tends to be smaller than exponential

Table 4.4 Comparison of basic wind speed and power law exponent

Route	Kobe-Naruto		Kojima-Sakaide	Onomichi-Imabari			
Name of bridge	Akashi-Kaikyo[a]	Onaruto[b]	All bridges[b]	Shin-Onomichi[c]	Tatara[c]	Kurushima-Kaikyo[c]	Others[b]
Basic wind speed	46 m/s	50 m/s	43 m/s	30 m/s	37 m/s	40 m/s	37 m/s
Power law exponent	1/8	1/7	1/7	1/5.5	1/7	1/7	1/7

[a]Wind Resistant Specification for Akashi-Kaikyo Bridge (1990)
[b]Wind Resistant Design Standard for HSB (1976)
[c]Wind Resistant Design Standard for Onomichi-Imabari Route (1994)

approximation in lower frequency range when two points are located with a distance such as 42 m.

In fact, when the buffeting prediction of the full bridge model of the Akashi-Kaikyo Bridge was made, the results were found substantially greater than the experimental results and this difference was attributed to the above difference between Davenport model and the field observation. Because the spatial correlation is lower than the assumed exponential model in low frequency range, such difference would become more significant for very long-span bridges with main span of 2,000 m or so. In such cases, von Kármán's model or something similar to it would be more appropriate. Effects of such modeling have been studied [3].

4.3 Wind Resistant Design of Honshu-Shikoku Bridges

4.3.1 Wind Resistant Design Standard for HSB (1976)

The *Wind Resistant Design Standard* (*1976*) [14] was the first standard applied to the design in reality, as is shown in the chronological Table 3.1. The basic wind speeds in this standard and other specification/standard are listed in Table 4.4 together with the power law exponents. Basic wind speeds are given as 10-min average wind speed at 10 m above the water level at the site. The return period for deciding the basic wind speed was 150 years, assuming that the expected life span of the bridge was 100 years and the probability of not exceeding the given threshold in this period was 50%.

Some of the background information of the standard is given in the following sections.

4.3.1.1 Basic Wind Speeds

The basic wind speeds over the sea at the bridge site were estimated by applying a regression analysis using the topographic factors. Based on the expected extreme wind

Table 4.5 Factor ν_2 of stiffening girder of suspension bridge

Height (m) \ Span length (m)	500	650	800	1,000	1,200	1,500	1,800
20–80	1.17	1.16	1.15	1.14	1.13	1.12	1.12

speeds with the 150 year return period at 22 meteorological observatories and weather stations with 26 topographic factors, the regression equations for the bridge site were formulated first. Then valid factors were selected by multi-regression analysis. Based on this analysis and also considering various uncertainties in the procedure, the basic wind speeds were finally determined with some engineering judgment to some extent.

4.3.1.2 Power Law Exponents and Design Wind Speeds

Wind speed profile changes according to the topographic conditions of the terrain. For topography with homogeneous roughness, wind speed profile can be expressed by the power law. Power law exponent is related to the surface terrain and averaging time, and the standard value for the 10-min average wind speed was taken as 1/8 until the *Wind Resistant Design Standard* (*1972*). However, in this *Standard* (*1976*), the power law exponent was set to 1/7, referring to the newly measured strong wind data at seacoasts and also by referring to some wind codes in foreign countries.

The design wind speed, U_D, is provided in this *Standard* (*1976*) as (4.25).

$$U_D = \nu_1 \cdot \nu_2 \cdot U_{10} \text{ (Horizontally stretched structures)}$$
$$U_D = \nu_1 \cdot \nu_3 \cdot U_{10} \text{ (Vertically stretched structures)} \tag{4.25}$$

ν_1 is a factor for the correction based on the representative height of the structure. The vertical wind speed profile is considered by power law in this factor. However, if wind speed increase is caused by surrounding terrain like a cape or peninsula, such effects should be added.

The effects of the wind speed fluctuation are included in U_D as factors ν_2 and ν_3. These factors are determined to give equivalent uniform wind load as the peak wind load under fluctuating natural wind. Effects such as aerodynamic admittance and mechanical admittance are not included in these factors, and they are considered in the expression of design wind load. The reason why different factors for horizontally and vertically stretched structures, ν_2 and ν_3, respectively, are defined is the difference of the spatial correlation of natural wind in the separation of horizontal and vertical direction. Exponential expression was used for the spatial correlation and exponents for horizontal and vertical direction separation were assumed to be 7 and 8, respectively. Example values of ν_2 computed for a stiffening girder of a suspension bridge are shown in Table 4.5.

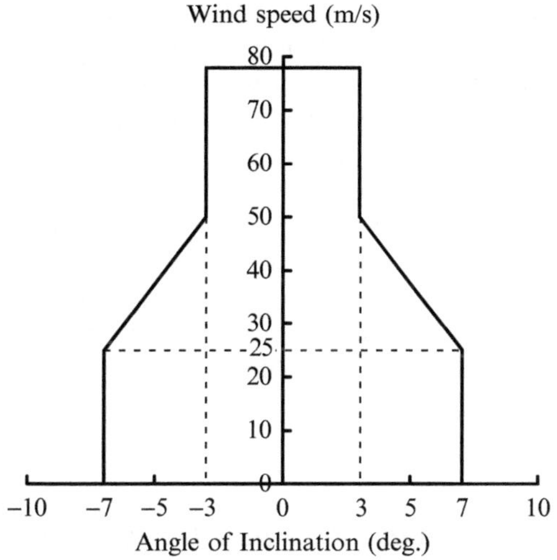

Fig. 4.10 Reference wind speed as a function of inclination angle

4.3.1.3 Reference Wind Speed for Self-Excited Vibration

In this *Standard* (*1976*), the critical wind speed for self-excited vibration is expected to be higher than 1.2 times design wind speed (4.26). The safety factor of 1.2 was determined considering (1) the possible errors in the wind tunnel tests, design and construction, and (2) the importance of the bridge structure together with the catastrophic nature of the phenomena. This threshold is a reference wind speed for the self-excited vibration.

$$1.2 \times U_D \tag{4.26}$$

4.3.1.4 Angle of Inclination

The effect of velocity fluctuation is partly taken into account by considering the inclination of the mean wind, with the condition that there is no bias in real mean inclination angle at the construction site. In this standard, the maximum inclination angle 7° was determined as three times of average standard deviation of inclination angle with averaging time of 30 s. The standard reference wind speed is shown in Fig. 4.10 as a function of inclination angle.

In addition, if there is a bias in mean inclination angle due to surrounding terrain effects, it has to be considered properly.

4.3.1.5 Basic Wind Speed During the Erection Period

Assuming the erection period of 2 years and taking the probability of not exceeding the threshold wind speed in this period to be 60%, the return period becomes approximately 5 years. The ratio of wind speed for 5 years to that of 150 years, R_5/R_{150}, is 0.66 based on the measured wind at observatories in the Seto Inland Sea area. Thus, the ratio of wind loading during the construction period to the service period becomes $0.66^2 = 0.436$. From this calculation, the standard wind load during erection was set as a half of that for the design of completed stage, giving a small safety margin.

However, there were cases where the maximum wind speed recorded in the past at or near the sites concerned were used to check the safety of bridges during the erection. These were the cases when the bridges were extremely sensitive to wind during erection, or when the consequence of wind-induced damage during the erection was unacceptably large.

4.3.2 *Wind Resistant Design Specification for Akashi-Kaikyo Bridge (1990)*

4.3.2.1 Basic Wind Speed

Wind speed data at the Tarumi observatory near the bridge site were continuously obtained for over 20 years since 1964, and it was considered to be the most reliable wind data for the design of the Akashi-Kaikyo Bridge. The basic wind speed was calculated based upon that wind speed data.

First, an expected wind speed for a return period of 150 years was calculated to be 49.4 m/s, considering the correction of different types of anemometers employed during the observation period. However, the obtained maximum wind speed data for each year did not fit well to the statistical extreme distribution of type I probably due to lack of statistical uniformity of the data. By considering these statistical uncertainties, the basic wind speed was determined by multiplying 1.1 so that the return period of 150 years can be ensured.

$$49.4 \times 1.1 = 54.3\text{m/s} \tag{4.27}$$

In order to find the wind speed ratio at the observatory to the bridge deck level (80 m above sea level), the wind tunnel test was carried out using a topographic model of the bridge site. From the results of the wind tunnel test (Fig. 4.11), it was confirmed that the wind speed ratio was between 1.06 and 1.10. Therefore, using the power law exponent of 1/8 (Table 4.4) for vertical wind speed profile, the basic wind speed at 10 m above sea level was determined as 46 m/s.

$$49.4 \times 1.1 \times 1.1 \times (10/80)^{(1/8)} = 46\text{m/s} \tag{4.28}$$

Fig. 4.11 Wind tunnel test using a topographic model of the Akashi-Kaikyo Bridge site (model scale: 1/5,000)

4.3.2.2 Power Law Exponent and Design Wind Speed

The value of 1/8 for the power law exponent of the vertical wind speed profile near the seacoast was deemed appropriate, considering the observed wind speed data at the Tarumi, Ohnaruto, and other bridge sites.

In this *Specification* (*1990*) [15], the design wind speed is to be calculated considering only the vertical wind speed profile as (4.29). The modification factor v_1 is the same as v_1 in Sect. 4.3.1.2.

$$\bar{U}_z = \mu_1 \cdot U_{10} = \left(\frac{z}{10}\right)^{0.125} \cdot \bar{U}_{10} \tag{4.29}$$

4.3.2.3 Reference Wind Speed for the Self-Excited Vibration

In this *Specification* (*1990*), the critical wind speed for self-excited vibration is expected to be higher than the threshold value of (4.30). $\bar{U}_z$ is the design wind speed of suspended stiffening girder at the representative height of z. μ_F is a factor that considers wind speed fluctuation.

μ_F is a ratio of averaged maximum wind speed in 10 min to the mean wind speed, $\bar{U}_z$. The averaged maximum wind speed is calculated with averaging time τ over the span length, where τ is the time needed for the occurrence of self-excited vibration. τ was assumed to be 30 s. Then by multiplying μ_F to $\bar{U}_z$, the maximum wind speed based on $\bar{U}_z$ that possibly onset the self-excited vibration is obtained.

Compared with the expression of (4.26), μ_F is necessary in (4.30) because the effects of wind speed fluctuation are not included in $\bar{U}_z$. With span length of 2,000 m, μ_F was calculated to be 1.08.

Also the safety factor of 1.2 was adopted as in Sect. 4.3.1.3. The value in (4.30) is a reference wind speed for the self-excited vibration.

$$1.2 \times \mu_F \times \bar{U}_Z = 1.2 \times 1.08 \times \bar{U}_Z \tag{4.30}$$

4.3.2.4 Angle of Inclination

Because of the flat terrain on the site and small static torsional deformation characteristics of the bridge in strong wind, the angle of inclination was set as in Table 3.5 in this *Specification* (*1990*). Effects of fluctuating inclination was also considered there based on a research [16], where the maximum angle of inclination averaged over the time of vibration development was taken into account. Some discussions on angle of attack are given in Sect. 6.4.2.

4.3.2.5 Basic Wind Speed During Erection Period

Assuming that the erection period is 5 years and non-exceeding probability is 0.8 during the period, return period is calculated to be 22.9 years. From the result of wind speed observation at Tarumi, the following value was set as the basic wind speed during erection period.

$$U_{22.9}/U_{150} = 0.81 \tag{4.31}$$

$$46\text{m/s} \times 0.81 = 37\text{m/s} \tag{4.32}$$

4.3.3 Wind Resistant Design Standard for Onomichi-Imabari Route (1994)

The content of this *Standard* for the Kurushima-Kaikyo Bridges and the Tatara Bridge [17] is given in Sect. 3.1.5. On top of these bridges, this *Standard* was also applied to the design of the Shin-Onomichi Bridge.

Table 4.6 Modification factor for verification of self-excited vibration

Bridge name	Shin-Onomichi Bridge	Tatara Bridge	Kurushima Bridges
μ_F	1.2	1.1	1.1

4.3.3.1 Basic Wind Speed

As for the Kurushima-Kaikyo Bridges and the Tatara Bridge, the basic wind speed was adopted from the 1976 code, since there were no long-term wind data at the bridge sites since 1976. On the other hand, the basic wind speed for the Shin-Onomichi Bridge was newly defined, considering the correlation between the wind data observed at the tower top and at the meteorological observatory which is closely located.

4.3.3.2 Power Law Exponent and Design Wind Speed

Power law exponents for the Kurushima Bridges and the Tatara Bridge were decided in the similar manner as that of the former specification, because their construction sites were characterized as an archipelago area, as opposed to 1/8 for the Akashi-Kaikyo Bridge that is located in an open sea area. On the other hand, in case of the Shin-Onomichi Bridge, the power law exponent was determined considering the hilly topography of the area. Based upon the observed turbulence intensity of 17% at the site, the α-value was decided as 1/5.5, considering the relationship between wind speed profile and turbulence intensity indicated in the *Wind Resistant Design Manual for Highway Bridges* [18].

4.3.3.3 Reference Wind Speed for Self-Excited Vibration

In the same way as the case of the Akashi-Kaikyo Bridge, the reference wind speed is given as the following expression

$$1.2 \times \mu_F \times \bar{U}_Z \tag{4.33}$$

The value of μ_F was set as shown in Table 4.6 for each bridge.

4.3.3.4 Angle of Inclination

Angle of inclination for the Tatara Bridge and Kurushima-Kaikyo Bridges should be set as in Table 3.5 in this *Standard*. It is the same as for the Akashi-Kaikyo Bridge (Sect. 4.3.2.4) because the effects of the surrounding terrain on the angle of inclination were found to be small from wind tunnel tests employing the topographic models.

Table 4.7 Parameters of wind characteristics

Items	I	II	III	IV
Displacement height of zero-plane above ground z_b (m)	5	10	15	30
Power law exponent α	0.12	0.16	0.22	0.29
Gradient height z_G (m)	500	600	700	700
Roughness length z_0 (m)	0.01	0.05	0.3	1.0

4.3.3.5 Basic Wind Speed During the Erection Period

Assuming that the erection period would be 4 years for Kurushima and Tatara and 2 years for the Shin-Onomichi Bridge, and that the probability of not exceeding the given threshold is 0.8 during these periods, the return period was calculated to be 18.9 and 9.5 years for each bridge. Based on this result, the basic wind speeds for the erection period became 31 m/s for Kurushima, 29 m/s for Tatara and 21 m/s for the Shin-Onomichi Bridge.

4.4 Wind Resistant Design Manual for Highway Bridges (1991)

4.4.1 Basic Wind Speed

Basic wind speed U_{10} is defined as the 10 min mean wind speed at 10 m height. The return period of the basic wind speed is considered to be 100 years, which means that the probability of not exceeding this wind speed during the bridge's lifetime, usually 50 years, is 60%. For the extreme value statistical analysis, Gringorten's momentum method was employed using the wind speed data obtained at the regional meteorological observatories all over Japan, assuming that Type I extreme value statistical model can be applied. Basic wind speed is then modified to the value corresponding to that of the surface terrain category II (countryside), considering the surface terrain at each observatory (Fig. 2.2) [18].

4.4.2 Power Law Exponent and Design Wind Speed

The factor E_I multiplied on the basic wind speed and turbulence intensity I_u are listed in Tables 2.7 and 2.8, respectively. These factors are calculated from the parameters of wind characteristics given in Table 4.7.

Table 4.8 Characteristics of fluctuating wind speed

	Turbulence intensity	Scale of turbulence	Shape of spectra	Spatial correlation
To be used for buffeting analysis	$\begin{cases} I_u = \frac{1}{\ln(30/z_0)}\left(\frac{30}{z}\right)^{\alpha} \\ (z_b<z<100m) \\ I_u = \frac{1}{\ln(30/z_0)}\left(\frac{30}{z_b}\right)^{\alpha} \\ (z \leq z_b) \end{cases}$ $I_v = 0.88 \cdot I_u$ $I_w = 0.5 \cdot I_u$	Reference values are prescribed	Kármán's model	Exponential function (k = 8)

$$\frac{U_d}{U_G} = \begin{cases} \left(\frac{z_b}{z_G}\right)^{\alpha}, & z \leq z_b \\ \left(\frac{z}{z_G}\right)^{\alpha}, & z_b<z \leq z_G \\ 1 \quad , & z_G<z \end{cases} \tag{4.34}$$

where, U_d is the basic design wind speed, z_b is the displacement height of zero-plane above ground and U_G is the wind speed at the gradient height, respectively. U_G can be calculated from above-mentioned basic wind speed.

$$U_G = \left(\frac{600}{10}\right)^{0.16} \cdot U_{10} \tag{4.35}$$

4.4.3 *Characteristics of Fluctuating Wind Speed*

Characteristics of fluctuating wind speed were prescribed as in Table 4.8 [19, 20].

4.4.4 *Reference Wind Speed for Self-Excited Vibration*

According to the *Wind Resistant Design Standard for HSB* (*1976*), the reference wind speed for self-excited vibration is given by the basic design wind speed multiplied by the factor E_{rl}, based upon the effect of fluctuating wind speed, and the safety factor of 1.2.

$$U_{rd} = 1.2E_{rl} \cdot U_d \tag{4.36}$$

The factor E_{rl} is given in Table 2.10, in accordance with the surface terrain category of the construction site. However, E_{rl} is set to be 1.0 for the examination of galloping, because the onset wind speed of galloping of a girder and tower becomes generally higher in turbulent flow than in uniform flow.

4.4.5 Angle of Inclination

For the examination of flutter in smooth flow, angle of inclination is basically set between $-3°$ and $+3°$ as in *Wind Resistant Design Specification for Akashi-Kaikyo Bridge (1990)* [15]. But for galloping, it is basically set between $-1°$ and $+1°$, because galloping occurrence is less likely in highly turbulent wind, and also the torsional deformation caused by wind is often negligible for a bridge that may suffer galloping. For vortex-induced vibration, angle of inclination is basically set between $-3°$ and $+3°$. This is determined based on a kind of balance of two: (1) the wind speed for examination of vortex-induced vibration is relatively low and the static torsional deformation is supposed to be small, and (2) with lower wind speed, turbulence intensity tends to be high.

In all cases, effects of surrounding terrain on the mean angle of inclination of wind have to be considered as well.

4.4.6 Basic Wind Speed During the Erection Period

If the non-exceeding probability during the erection period is set, corresponding return period, R, can be obtained. The non-exceeding probability is often set to be about 0.6 by taking the same value for a completed bridge in its service life.

It is preferable that the relationship between the return period R and the expected wind speed is determined based upon the observed wind data at the construction site. But if it is difficult, the basic wind speed for the erection period U_{10E} can also be determined by (4.37) using the basic wind speed U_{10}, that is obtained as the expected maximum wind speed for the return period of 100 years. Equation (4.37) is derived by assuming the Gumbel distribution for the yearly maximum wind speed distribution and the average parameters based on the data at the meteorological observatories in Japan.

$$U_{10E} = U_{10}\left\{0.61 - 0.10\ln\left(\ln\frac{R}{R-1}\right)\right\}/1.07 \tag{4.37}$$

4.5 Examples of Wind Characteristic Prediction Considering Topography

Usually, the effects of topography on the design wind characteristics are not taken into consideration when the construction sites are relatively flat, e.g., any straits, plain, and urban area. Wind resistant design can be carried out using various wind

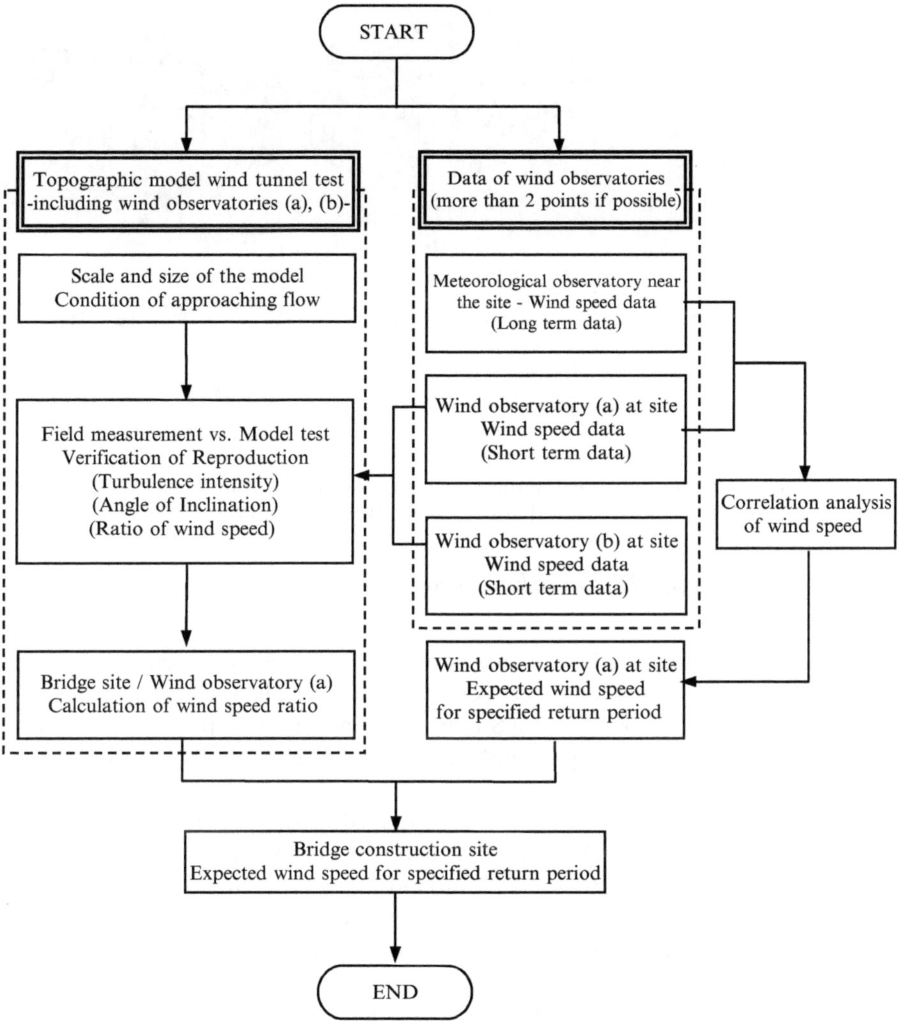

Fig. 4.12 An example of flowchart using the measured wind data and wind tunnel tests with topographical models

characteristics specified by above-mentioned codes or standards in these cases. On the other hand, the effect of topography on the design wind characteristics must be considered in case the non-uniformity of the terrain is significant, e.g., straits with sparsely scattered islands, straits with steep escarpments on the sides, a mouth of gulf, and ravines, etc.

Currently, the method combining the measured wind data near the sites and wind tunnel tests with the topographic models is mainly used in this procedure. An example is shown in Fig. 4.12.

A couple of sample bridges to which the above-mentioned method was applied are shown in the following.

Fig. 4.13 Full aeroelastic model and surrounding topography for the Kurushima Bridges with the model scale of 1/150 (courtesy of the Honshu-Shikoku Bridge Authority)

4.5.1 The Kurushima-Kaikyo Bridges

The Kurushima-Kaikyo Bridges consists of three consecutive suspension bridges crossing the Kurushima Straits between Oshima and Shikoku Islands. There are many islands around the Kurushima Straits, and hence it was expected that the wind characteristics would be largely influenced by the existence of these islands, depending on the wind direction.

First, the wind tunnel tests with a topographical model of the area were carried out in order to investigate the wind characteristics at the bridge site. The model reproduced the area of 9 km diameter around the bridge site with 1:1,500 linear scale. Later, a large wind tunnel with the test section of 41 m wide and 4 m high was employed for more detailed tests, using the full aeroelastic bridge model together with the surrounding topography with 1:150 linear scale (Fig. 4.13). The aerodynamic stability considering the effect of topography was thus investigated.

The map of the bridge site and the resulted wind characteristics are shown in Figs. 4.14 and 4.15, respectively. It was found that the effect of the islands closely located in the windward side was particularly significant. The mean wind speed for this case was reduced as a result and turbulence intensity was increased [21, 22].

The basic wind speed for the Kurushima-Kaikyo Bridges was specified in the standard as mentioned, but turbulence intensity and the angle of inclination measured in the tests formed the basis for another wind tunnel test that used a full aeroelastic model and the surrounding topography.

4.5.2 The Megami Bridge

The Megami Bridge is a 3-span continuous steel cable-stayed bridge with the center span of 480 m. The bridge is located at the mouth of the Nagasaki Bay,

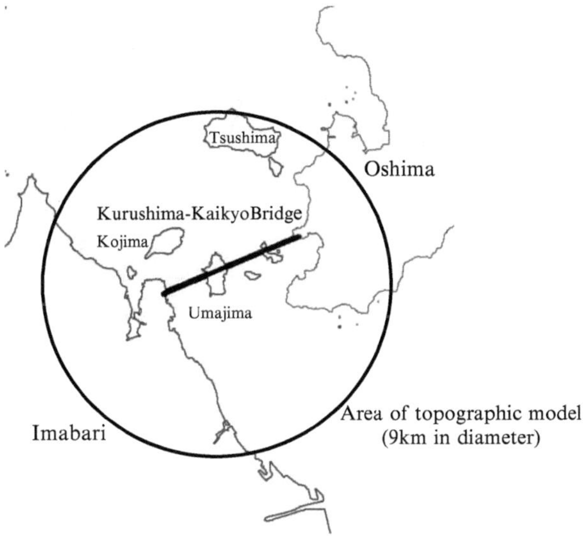

Fig. 4.14 Construction site of the Kurushima-Kaikyo Bridges

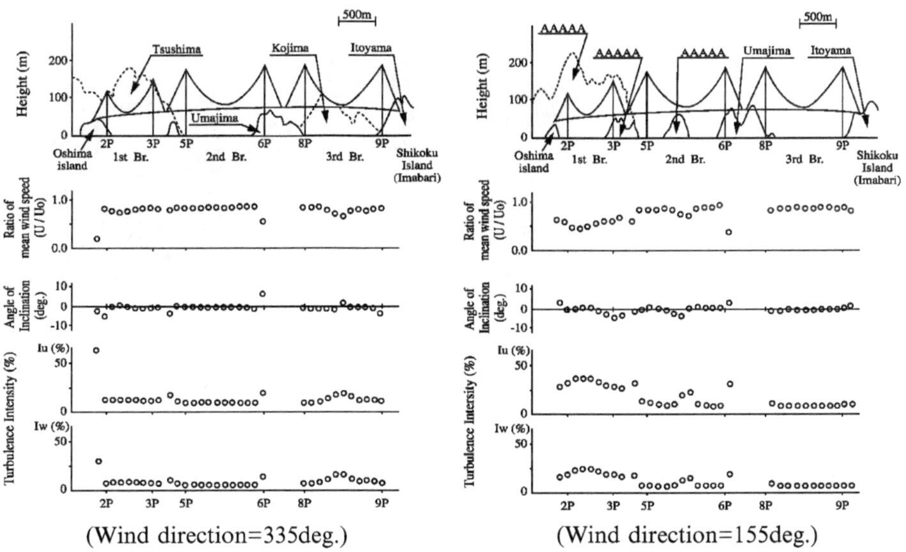

Fig. 4.15 Results of the wind tunnel tests using the topographical model (the Kurushima-Kaikyo Bridges) [21, 22]

where the hilly coastlines are making a narrow channel (Fig. 4.16). The aerodynamic stability of the bridge was considered to be strongly influenced by the wind turbulence because of this topography. Wind tunnel tests with a 1:2,000 scale topographical model corresponding to a 12 km diameter area around the bridge

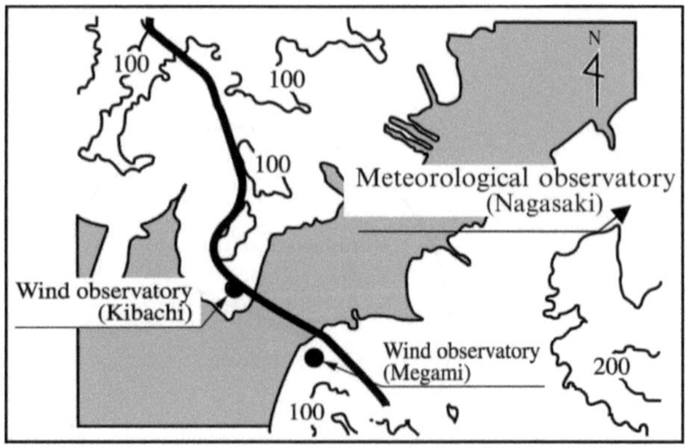

Fig. 4.16 Construction site of the Megami Bridge

Fig. 4.17 Topographical model for the Megami Bridge (1/2,000, wind dir: SSW)

site was carried out in order to investigate the wind characteristics at the site (Fig. 4.17).

The expected wind speed at wind observatories for 100-year return period for each wind direction was calculated from the wind speed correlation analysis. The results are given in Fig. 4.18 in comparison with the measured data at the Nagasaki meteorological observatory.

The comparison between the wind tunnel test results and the observed data are shown in Fig. 4.19. The wind tunnel test results show generally good agreement with the onsite measurement. Thus it was concluded that the wind characteristics at the bridge site could be predicted by the use of the wind tunnel tests with topographical models [23–25].

Then the design wind speed at the bridge site was calculated as shown in Fig. 4.12, using the wind speed ratio (Bridge site/Kibachi observatory) obtained

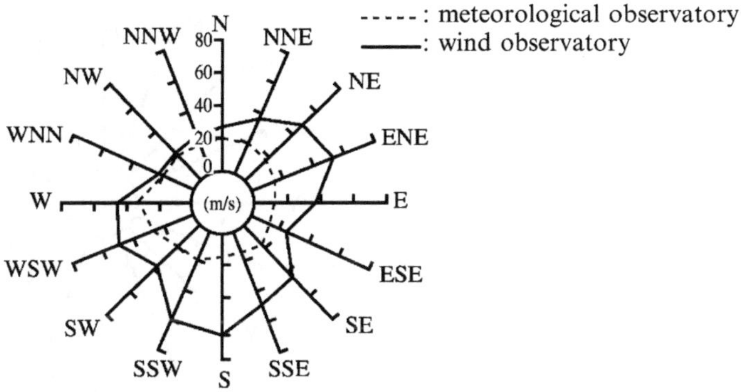

Fig. 4.18 Expected wind speed for $R = 100$ years at the Kibachi observatory

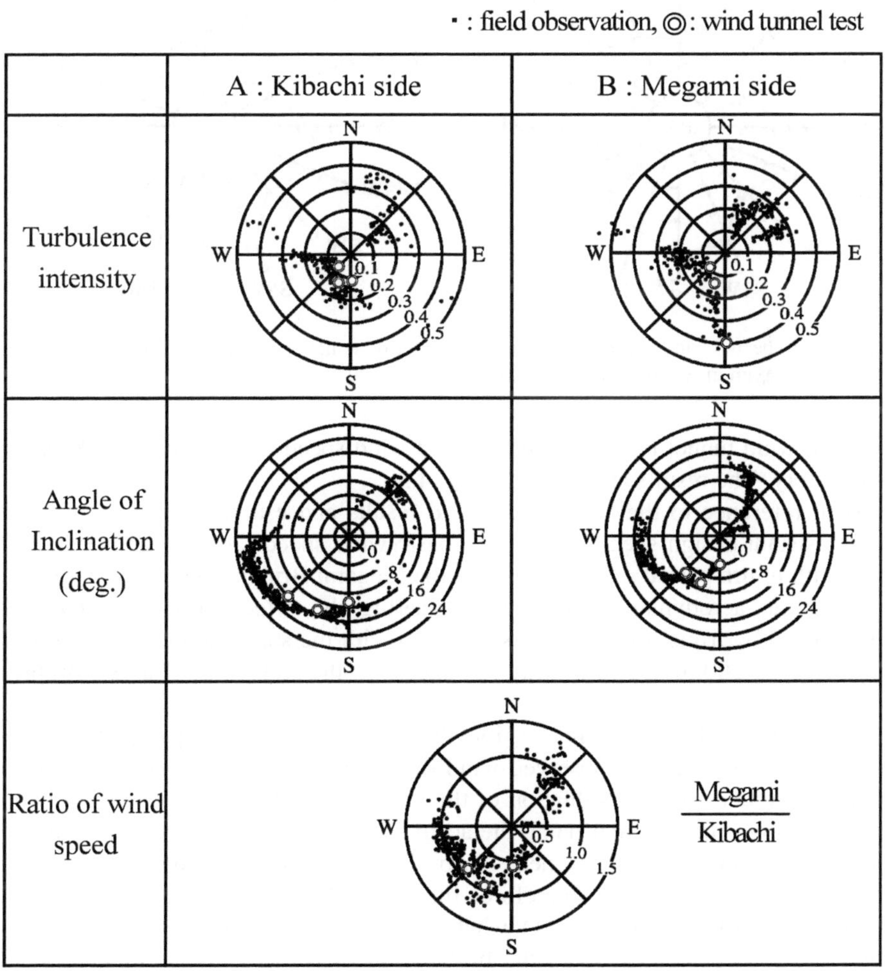

Fig. 4.19 Comparison of wind characteristics between the result of wind tunnel test and the observed data

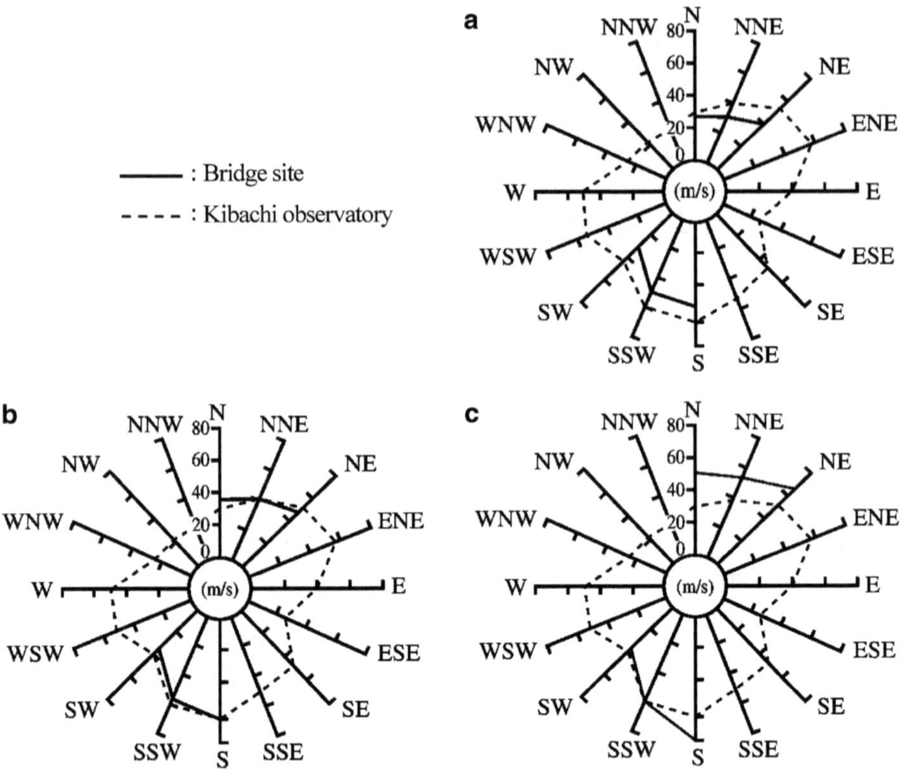

Fig. 4.20 Expected wind speed of 100-year return period at the bridge site. (**a**) Mid span at 10 m height. (**b**) Maximum value at the girder. (**c**) Gradient wind speed

from the wind tunnel tests and the expected wind speed calculated at the Kibachi observatory. The results are shown in Fig. 4.20.

With reference to the above-mentioned results and estimated values by other wind resistant design specifications, the basic wind speed (at 10 m above the sea level with return period of 100 years) for the Megami Bridge was set to be 49 m/s.

Also, the aerodynamic stability considering the effect of topography was investigated by using a full aeroelastic model with the surrounding topography at the scale of 1:200, in the same way as the case of the Kurushima-Kaikyo Bridges (Fig. 4.21).

As summarized in the former section, the combination of the topographic model wind tunnel tests and the observed wind data near the bridge site consists of an effective method for the evaluation of the topography effects on the design wind characteristics. On the other hand, the accuracy of expected wind speed measurement at the wind observatory near the bridge site is an important factor if this method would be applied. Therefore, further studies are required to solve the remaining problems such as: (1) what could we do when the meteorological observatory is located far from the bridge site; or (2) what if the data of wind

Fig. 4.21 Full aeroelastic model and surrounding topography for the Megami Bridge (model scale: 1/200)

direction observed at the meteorological observatory are very different from those of the wind observatory near the bridge site.

References

1. Tokoro S, Matsunaga I, Yoshitake N, Honda A, Kameda Y (1998) A study on the aerodynamic stability of Shin-Kitakyushu Airport Access Bridge using full aeroelasic model (in Japanese). In: Proceedings of the 15th national symposium on wind engineering, pp 521–526
2. Katsuchi H, Yamada H, Miyata T, Tanaka H, Ito S (2000) Wind properties on Ohnaruto bridge and Akashi-Kaikyo bridge (in Japanese). In: Proceedings of the 16th national symposium on wind engineering, pp 47–52
3. Toriumi R, Katsuchi H, Furuya N (2000) A study on spatial correlation of natural wind. J Wind Eng Indust Aero 87:203–216
4. Hino M (1968) Maximum ground-level concentration and sampling time. Int J Atmos Environ 2:149–165
5. Hino M (1971) Spectrum of gusty wind. In: Proceedings of the 3rd international conference on wind effects on buildings and structures, Tokyo, pp 1–7
6. von Kármán T (1948) Progress in the statistical theory of turbulence. Proc Natl Acad Sci USA 34:530–539
7. Harris RI (1968) On the spectrum and auto-correlation function of gustiness in high winds. Electrical Research Association, Report, 5273
8. Busch N, Panofsky H (1968) Recent spectra of atmospheric turbulence. Q J Roy Meteorol Soc 94:132–148
9. Shiotani M, Arai H (1970) Observation of strong wind at Nakagawa river in Tokushima (in Japanese). In: Proceedings of the 1st national symposium on wind engineering, pp 57–64

10. Honshu Sikoku Bridge Authority and Kansai Headquarters of Japan Weather Association (1985) 20-year wind database of Tarumi observation mast (in Japanese). HSBA, Kobe
11. Kansai International Airport Corporation and Japan Institute of Construction Engineering (1986) Report on the superstructure of Kansai international airport access bridge (in Japanese)
12. Honshu-Shikoku Bridge Authority (1986) Report of the analysis of strong wind data at Ohnaruto bridge site (in Japanese). HSBA, Kobe
13. Davenport AG (1961) The Spectrum of horizontal gustiness near the ground in high winds. Q J Roy Meteorol Soc 87:194–211
14. Honshu-Shikoku Bridge Authority (1976) Wind resistant design standard (in Japanese). HSBA, Kobe
15. Honshu-Shikoku Bridge Authority (1990) Wind resistant design specification for Akashi-Kaikyo bridge (in Japanese). HSBA, Kobe
16. Miyata T (1982) Evaluation of inclination due to wind speed fluctuation for bridge deck instability (in Japanese). J Wind Eng JAWE (13):29–35
17. Honshu-Shikoku Bridge Authority (1994) Wind resistant design standard for Onomichi-Imabari route (in Japanese). HSBA, Kobe
18. Japan Road Association (1991) Wind resistant design manual for highway bridges (in Japanese). Maruzen, Tokyo. Also revised version in 2007
19. Davenport AG (1982) The interaction of wind and structures. In: Plate EJ (ed) Engineering meteorology. Elsevier, Amsterdam, pp 527–572
20. ESDU (1974) Characteristics of atmospheric turbulence near the ground. Engineering Science Data Unit Ltd, Item No. 74031
21. Saeki S, Okukawa A, Ohashi H (1998) The Honshu-Shikoku Bridge. Struct Eng Int IABSE 8(1):10–15
22. Hirai S, Ohashi H, Kusuhara S, Honda A (1992) Experimental study on characteristics of natural wind over the Kurushima straits (in Japanese). In: Proceedings of the 12th national symposium on wind engineering, Tokyo, pp 25–30
23. Honda A, Fukahori S, Imagane S, Ishioka F (1999) Aerodynamic design of cable-stayed bridges surrounded by undulated topography. In: Proceedings of the tenth international conference on wind engineering, Copenhagen, pp 913–918
24. Honda A, Miyazaki S, Ishioka F, Sakaniwa Y (1994) Wind resistant design for open-to-sea, hill-flanked Megami bridge (in Japanese). In: Proceedings of the 13th national symposium on wind engineering, Tokyo, pp 479–484
25. Honda A, Fukahori S, Imagane S (1998) Study on the topographic effects to the gust response of long spanned cable stayed bridge (in Japanese). In: Proceedings of the 15th national symposium on wind engineering, Tokyo, pp 383–388

Chapter 5
Structural Damping

5.1 Introduction

Structural damping of bridges needs to be specified in a wind resistant design code for examining their dynamic behavior against anticipated wind action. Because the dynamic response of bridges is generally evaluated by using the modal analysis, the quantity of structural damping has to be represented in terms of modal damping. The damping specification in the design code, therefore, is given in modal damping, based on the results of field measurements. The design and construction of long span bridges, such as Honshu-Shikoku bridges, have been continuously going on in Japan since 1970s. The modal damping data of long span bridges have been accumulated throughout this period by conducting field vibration tests, often just after completion of their construction, in order to verify the design assumption against the reality. At the same time, the modal damping data of medium span bridges have been also measured and accumulated by field vibration tests, since the field measurements are not only for the wind resistant design but also for the consideration of vehicle-induced dynamic responses. The structural damping values in logarithmic decrement are specified in wind resistant design codes by referring to these measured data, with the addition of some engineering judgments, since the theoretical evaluation of structural damping is very difficult at present.

5.2 Field Measurement of Structural Damping by Vibration Tests

5.2.1 Vibration Testing Methods for Bridges

There are several testing methods for measuring the structural damping of bridges, among which the shaker-based dynamic tests and the ambient vibration measurements are summarized here [1].

Y. Fujino et al., *Wind Resistant Design of Bridges in Japan*,
DOI 10.1007/978-4-431-54046-5_5, © Springer 2012

5.2.1.1 Forced Vibration Tests Using Shakers

There are two methods in the evaluation of structural modal damping by vibration tests using shakers: the free vibration method and forced vibration method. By using the free vibration method, the logarithmic decrement is directly calculated from the damped free vibration record. In this case, a shaker is used for inducing a resonant response of the bridge by exciting the structure at one of the natural frequencies, and the time histories of decaying free vibration is recorded after the shaker is stopped. In case of the forced vibration method, the modal damping is characterized in the resonance curve, or the frequency response function, by conducting a so-called sweep test. By applying a modal identification technique such as the half-power method, the magnitude of modal damping can be determined from the experimentally obtained resonance peaks.

The shaker-based vibration test is considered to be the best method for evaluating the structural damping in terms of data reliability, because the modal damping can be identified directly from a near-harmonic, single-mode response with relatively large amplitude. However, this method tends to be costly and has been used mainly for long span bridges in Japan [2, 3].

5.2.1.2 Ambient Vibration Tests

The identification of structural damping by ambient vibration surveys has been frequently carried out because of its practical easiness in application, and many of the damping data identified through this method are quoted for the damping specification in wind resistant design codes. As it has been often pointed out, however, the accuracy of structural damping identified by the use of the power spectral density is not necessarily very high and many studies to improve this precision have been conducted in relation to research demand for structural health monitoring. One of more accurate methods for identifying structural damping is the Random Decrement (RD) method, in which the damping is identified through constructing damped free vibration data from the gust response records by excluding uncorrelated components of externally excited responses. This method is accurate and capable of identifying the amplitude-dependent damping, while there still exist a problem of separating structural damping from the aerodynamic damping, since it is difficult to avoid the contribution of wind effects entirely. One example of RD method applications in Japan was to the Hakucho Bridge, a box-girder suspension bridge with the center span of 720 m, and the damping identification, including the aerodynamic damping, was successfully carried out [4]. More advanced method for damping evaluation is the Eigen-system Realization Algorithm (ERA) in the time domain, which has been successfully applied to some bridge structures. The damping measurement for the Hakucho Bridge and the Tatara Bridge has been reevaluated by applying the ERA method and found to produce results with better accuracy than before [5–7].

5.2.1.3 Other Methods of Damping Measurement

For medium span bridges, the shaker method and/or the multi-point measurement by ambient vibration survey are not practical because of their cost-performance. More convenient and less costly methods, such as the manual excitation method using cranes, with which an approximately periodic forced oscillation or an initial displacement to trigger the free vibration can be given, or the moving truck method and the impulse excitation method have been more frequently used. The accuracy of identified damping values, however, is generally lower than the tests using the shakers.

5.2.2 Shakers

The shakers used for field vibration tests of bridges, especially long-span bridges, in Japan are mainly of the following three types:

5.2.2.1 Shaker of Public Works Research Institute

This is an unbalanced-mass type shaker, as shown in Fig. 5.1a, and has been used since 1970s for the field vibration tests of bridges with widely ranged spans. By rotating eccentric masses attached to the rotors, the centrifugal forces are induced in the direction perpendicular to the rotor axis and used as the excitation force. There is, however, a difficulty in inducing large enough excitation force at low frequency by this method, because the force induced by eccentric mass is in proportion to the square of its frequency.

5.2.2.2 Shaker of Honshu-Shikoku Bridge Authority

The Honshu-Shikoku Bridge Authority (HSBA), in 1988, made two large shakers that induce large excitation forces by periodically driving up to 55 tons of weights with motors, gears and cranks, as depicted in Fig. 5.1b. The shakers have been used since then in field tests for the measurements of dynamic characteristics, including the structural damping [8]. As a background of these measurements, there was a need to measure the structural damping of long span bridges more accurately, particularly in vibration with large amplitude. This is because the logarithmic decrement of 0.03, specified in *Wind Resistant Design Standard for HSB* (*1976*), was actually the value based on the data measured on much shorter span bridges and also some results of long span bridges but in small amplitude range.

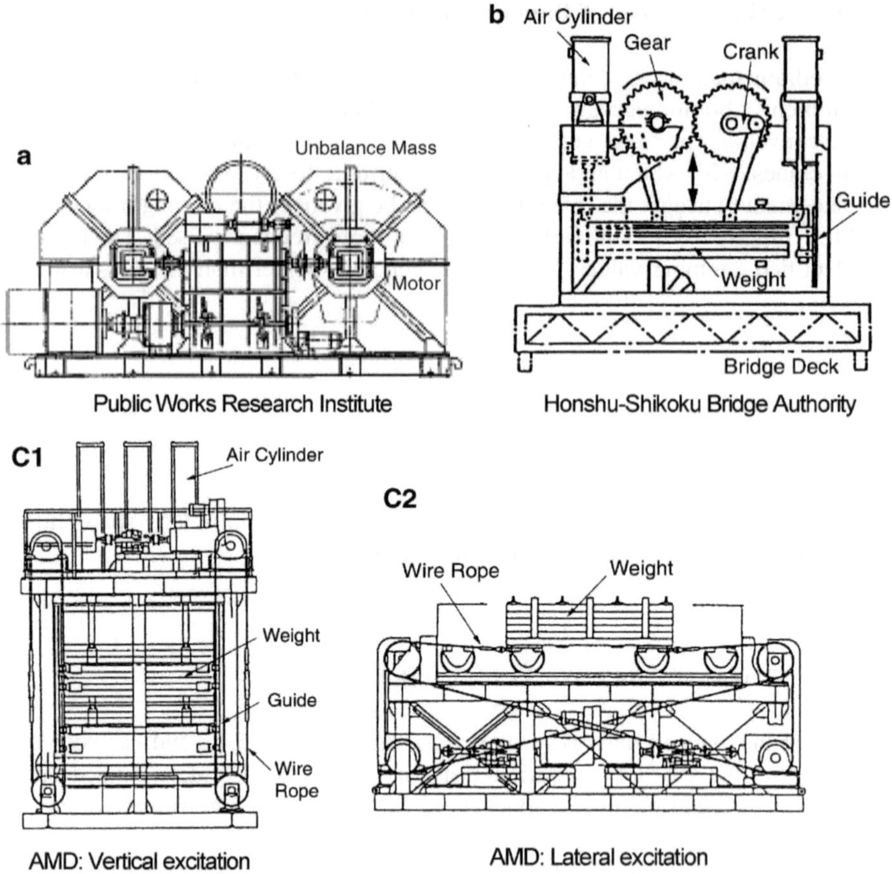

Fig. 5.1 Mechanical shakers for field vibration tests of long-span bridges in Japan: (**a**) PWRI, (**b**) HSBA, and (**c**) AMD

5.2.2.3 Active Mass Driver

Forced vibration method with active mass driver (AMD), which is an application of active-vibration-control technique, was adopted in the field vibration tests of the Tatara Bridge; a three-span composite box-girder cable-stayed bridge (270 + 890 + 320 m), in 1998 [9] and the Akinada Bridge; a three-span, two-hinged box girder suspension bridge (255 + 750 + 170 m), in 1999 [10]. The AMD can be easily handled because of its simple mechanism (Fig. 5.1c-1), and is a more convenient shaker that can induce the excitation forces in very low frequency range with very large stroke. The AMD can be used also as a lateral vibration shaker by simply rearranging the parts of AMD, as shown in Fig. 5.1c-2. In fact, by using this

Table 5.1 Specification of shakers

	Unbalance-mass type (Public Works Research Institute)	Gear-crank type (Honshu-Shikoku bridge authority)	AMD type (Mitsubishi heavy industry)
Exciting direction	Vertical and horizontal	Vertical	Vertical and horizontal
Total weight	12 tonf	90 tonf	37 tonf
Driving weight	Exciting force: 12 tonf (2 Hz)	33 tonf	20 tonf
Weight stroke		$\pm$ 0.50 m	$\pm$ 1.00 m
Exciting frequency	0.1–2.0 Hz	0.155–0.920 Hz	DC –
Operation system	In-phase or out-of-phase	In-phase or out-of-phase	In phase or out-of phase
Driving motor	Synchronous motor (37 kW)		AC servo motor (55 kW $\times$ 2)
Electric source	150 kVA		Generator (200 kVA $\times$ 2 set)

technique, the damping identification was done at very low frequency for the lateral vibration modes in the Tatara Bridge and the Akinada Bridge.

The specification of shakers of Public Works Research Institute, HSBA and the AMD for the Tatara Bridge, are summarized and compared with each other in Table 5.1.

5.2.3 *Methods of Measurements*

The measurement of modal structural damping is usually associated with that of the natural frequencies and the vibration mode shapes. This requires a multi-point measurement of vibration in the field experiment regardless of the method adopted; a vibration test with shaker or an ambient vibration test. Because of their reasonable cost, accelerometers are therefore frequently used as vibration sensors. It should be noted, however, that in the acceleration measurement, the higher frequency components tend to be overemphasized compared to the measurement of displacement, because the acceleration amplitude is proportional to the square of the frequency. The calculation of displacements by integrating the acceleration data would usually involve significant errors. In order to avoid these problems, displacements are sometimes directly measured by displacement-transducers such as optical sensors.

Figure 5.2 shows an example of a sensor system used for the shaker vibration test of the Tatara Bridge [7], where many accelerometers were placed even for the stay cables because of the possibility of their vibration coupled with girder motion. It is also noted here again that the damping identified from thus measured data includes aerodynamic damping to some extent and that the measurement of wind velocity by placing an anemometer is very important for evaluating the effect of aerodynamic damping.

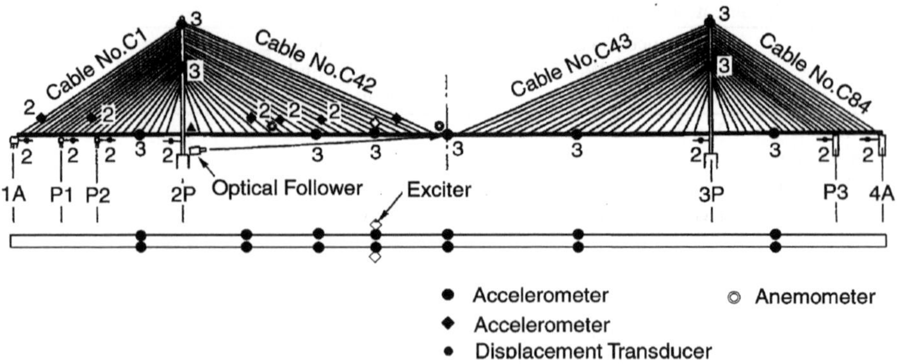

Fig. 5.2 Sensor placement in the Tatatra Bridge during the vibration test using the shaker

5.2.4 Characteristics of Available Damping-Data from Field Measurements

5.2.4.1 Data Inhomogeneity due to Difference of Testing Methods

Damping data from field measurements are not necessarily always directly comparable to each other because the amplitude level at which the damping was evaluated may be different for different measurements. The inhomogeneity and scatter in damping data caused by different levels of measurement accuracy and different test conditions, when they had to be considered as the background data for the design specifications, have been practically covered by introducing some engineering judgments.

5.2.4.2 Effects of Wind on Damping Evaluation

The vibration measurement is not necessarily always done under no wind conditions and there are some contribution of aerodynamic damping to measured damping data, which of course depends on the wind speed. This becomes a particularly sensitive issue for the case of stay cable vibrations, since their structural damping is extremely low. There have been some attempts to evaluate the structural damping by measuring the total damping at different levels of wind speed and extrapolate the results to zero-wind condition. However, in many cases, the contribution of aerodynamic damping is simply neglected by taking the measurement at low wind condition. References [5, 6] are a few examples that extract aerodynamic effects (damping and stiffness) on bridges from ambient motion under wind.

5.2.4.3 Effects of Coupled Cable Vibration in Cable-Stayed Bridges

Cable-stayed bridges are composed of many structural members such as the girder, pylons and cables, and the global modes of girder-dominant vibration can be coupled with local modes of other structural members such as cables. Especially in the case of long-span cable-stayed bridges whose natural frequencies in global modes become very small, there is a high possibility of these global modes coupled with the local cable vibrations. Effects of the coupled cable vibration on the structural damping have been studied [11, 12] quantitatively for the Ikuchi Bridge; a three-span composite box-girder cable-stayed bridge (150 + 490 + 150 m) [13], the Tsurumi-Tsubasa Bridge; a three-span steel box-girder cable-stayed bridge (255 + 510 + 255 m) [14, 15], and the Tatara Bridge [7, 16]. It should be noted that the structural damping of global modes of cable-stayed bridge can be made significantly smaller by having coupled with the local cable vibrations, because the cables generally have very small structural damping. However, in the case that some dampers are attached to cables for controlling the cable vibrations such as rain–wind vibration, the structural damping of global modes coupled with the local cable modes can be even larger because of large damping for this case through the controlled cable motion.

5.2.4.4 Effects of Friction

For the case of Tatara, the fairings are attached to the girder for aerodynamic stabilization. In its shaker-excitation tests, the structural damping in lateral bending mode of Tatara was found very large, and it was attributed to the friction due to sliding of fairings at their connection to the girder. It was indeed confirmed by measuring the sliding motion during the damping measurement [9]. This kind of friction between the structural parts, including the bearings and expansion joints, is one of the significant sources of structural damping. The contribution of friction should be carefully noted in the evaluation of measured damping data because the magnitude of friction damping often depends on the vibration amplitude.

5.3 Damping Data of Cable-Supported Bridges and Its Characteristics

5.3.1 Structural Damping of Girder-Dominant Mode

5.3.1.1 Suspension Bridges

Table 5.2 gives structural damping (in log-decrement) of girder dominant vibration modes, along with the corresponding natural frequencies, measured at some of the major suspension bridges in Japan. The damping values for the Minami Bisan-Seto,

Table 5.2 Natural frequencies and structural damping (logarithmic decrement) of girder-dominant modes in suspension bridges

Bridge name		Minami Bisan-Seto		Ohnaruto		Innoshima		Rainbow		Ohshima		Hakucho		Akinada	
Spans (m)		274 + 1100 + 274		93 + 330 + 876 + 330		250 + 770 + 250		114 + 570 + 114		140 + 560 + 140		330 + 720 + 330		255 + 750 + 170	
Vibration mode		Frequency (Hz)	Logarithmic decrement	Frequency (Hz)	Logarithmic decrement	Frequency (Hz)	Logarithmic decrement	Frequency (Hz)	Logarithmic decrement	Frequency (Hz)	Logarithmic decrement	Frequency (Hz)	Logarithmic decrement	Frequency (Hz)	Logarithmic decrement
Vertical	1st Asym.	0.151	0.088	0.167	0.108	–	0.024	0.255	0.15	0.189	0.018	0.149	0.016	–	–
	1st Sym.	0.168	0.037	0.166	0.108	0.178	0.035	0.260	0.06	0.232	0.017	0.129	0.034	–	–
	2nd Sym.	0.228	0.096	0.284	0.085	–	–	0.411	0.16	0.323	0.020	0.218	0.027	–	–
Torsion	1st Sym.	0.329	0.026	0.328	0.028	0.335	0.020	0.485	0.06	0.553	0.018	0.469	0.036	0.536	0.040
	1st Asym.	0.457	0.038	0.506	0.062	0.583	–	0.681	0.13	0.759	0.041	0.760	0.033	0.845	0.074
Lateral	1st Sym.	–	–	–	–	–	–	–	–	–	–	0.099	0.290	0.078	0.087
	1st Asym.	–	–	–	–	–	–	–	–	–	–	–	–	0.224	0.157
	2nd Sym.	–	–	–	–	–	–	–	–	–	–	0.560	0.039	0.466	0.066

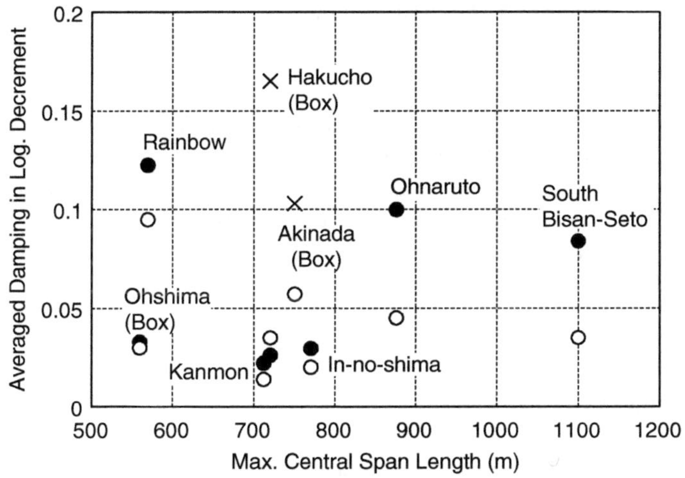

Fig. 5.3 Average damping of girder-dominant modes vs. max. span length in suspension bridges: (●) vertical mode, (○) torsional mode, and (×) lateral mode

Ohnaruto, and Ohshima Bridges were all identified by the vibration tests with the HSBA shaker. The HSBA shaker has been mostly used for the vibration tests of the Honshu-Shikoku bridges, but sometimes it was applied to other bridges such as the Rainbow Bridge [17], whose structural damping is also included in the table. It should be noted that the structural damping of the Innoshima Bridge was measured for small amplitude vibration in 1983 by the ambient vibration test and also by using the weight-dropping method [18], before the mechanical shaker of HSBA was manufactured in 1988.

Based on Table 5.2, the average structural damping of each suspension bridge was calculated for vertical bending and torsion, respectively. The results are plotted against the maximum span length of the bridge in Fig. 5.3. All the bridges given in Fig. 5.3 are suspension bridges with the span longer than 500 m. Note that the data for the Kanmon Bridge was obtained only for small amplitude vibration by using a smaller shaker of PWRI. Qualitative findings from Fig. 5.3 can be summarized as follows.

(a) Except for the Ohshima Bridge with a box girder, the average structural damping identified by the large shaker is relatively large, and decreases its value with the increase of the span length.
(b) For suspension bridges with a truss-stiffened girder, the structural damping in torsion is smaller than that of vertical bending.
(c) For suspension bridges with the box girder stiffening, the structural damping in both torsion and vertical bending are of the similar magnitude.
(d) The structural damping of suspension bridge is small with the box girder stiffening than with the truss-stiffening, when their span length is comparable.

Furthermore, regarding the first symmetric modes in vertical bending and torsional motions in Table 5.2, the following quantitative conclusions can be derived:

(e) For suspension bridges with a truss-stiffened girder, the structural damping (in log-decrement) of vertical bending motion is about 0.035 on average, while that of torsion ranges from 0.02 to 0.04. This is consistent with the damping value of 0.03 specified in the HSBA Code (1976).
(f) For suspension bridges with a box girder, the damping values are about 0.02 for both vertical bending and torsion.

5.3.1.2 Cable-Stayed Bridges

The results of field vibration tests using a shaker are summarized for cable-stayed bridges in Table 5.3, and the relation between the average structural damping and the maximum span length is depicted in Fig. 5.4 [16]. As it can be seen from Fig. 5.4, except for the Yokohama Bay Bridge [19], the Hitsuishijima Bridge [20], and the Rokko Bridge, whose girders are the truss type, and the Tsurumi-Tsubasa Bridge, in which artificial dampers are installed for controlling cable vibrations, the average structural damping of cable-stayed bridges decreases with the increase of maximum span length and seems to converge to 0.02 in log-decrement for longer spans in both vertical bending and torsion. On the other hand, the cable-stayed truss-girder bridges have relatively large structural damping of 0.07 or above.

5.3.1.3 Structural Damping for Girder-Dominant Modes Specified in Design Codes

Referring to the above discussion on the damping characteristics of the cable-supported bridges, the standard values of structural damping for girder-dominant modes are specified in the *Wind Resistant Design Standard for Honshu-Shikoku Bridges* (WRDSHSB (2001)), while the fundamental concept of damping specification in the previous *Wind Resistant Design Standard for HSB* (1976) is still being followed up. That is to say that the structural damping in torsion of truss-stiffened suspension bridge is specified as 0.02, different from that in vertical bending and is smaller than 0.03 specified for both vertical bending and torsion in the HSB Standard (1976). This is partly because the structural damping can be lower for longer span such as the Akashi-Kaikyo Bridge, even if the girder is truss-stiffened, and also partly because assuming a smaller structural damping is more conservative in design. As for the structural damping of both suspension and cable-stayed bridges with box girder stiffening, the log decrement of 0.02 is the standard for both vertical bending and torsion, because the frictional damping between members in the case of box-girder is expected to be less than truss-stiffening case.

Table 5.3 Natural frequencies and structural damping (logarithmic decrement) of girder-dominant modes in cable-stayed bridges

Bridge name		Tatara br.		Tsurumi-Tsubasa br.		Ikuchi br.		Yokohama bay br.		Hitsuishijima br.	
Spans (m)		270 + 890 + 320		255 + 510 + 255		150 + 490 + 150		200 + 460 + 200		185 + 420 + 185	
Vibration mode		Frequency (Hz)	Logarithmic decrement	Frequency (Hz)	Logarithmic decrement	Frequency (Hz)	Logarithmic decrement	Frequency (Hz)	Logarithmic decrement	Frequency (Hz)	Logarithmic decrement
Vertical	1st Sym.	0.226	0.024	0.213	0.07	0.340	0.018	0.351	0.069	0.430	0.080
	1st Asym.	0.263	0.018	0.293	0.04	0.443	0.020	0.565	0.134	0.729	0.088
	2nd Sym.	0.348	0.007	0.518	0.10	0.605	0.031	0.828	0.088	1.030	0.066
Torsion	1st Sym.	0.497	0.017	0.545	0.07	0.735	0.038	0.876	0.041	1.058	0.046
	1st Asym.	0.831	0.051	–	–	1.603	0.019	–	–	1.910	0.071

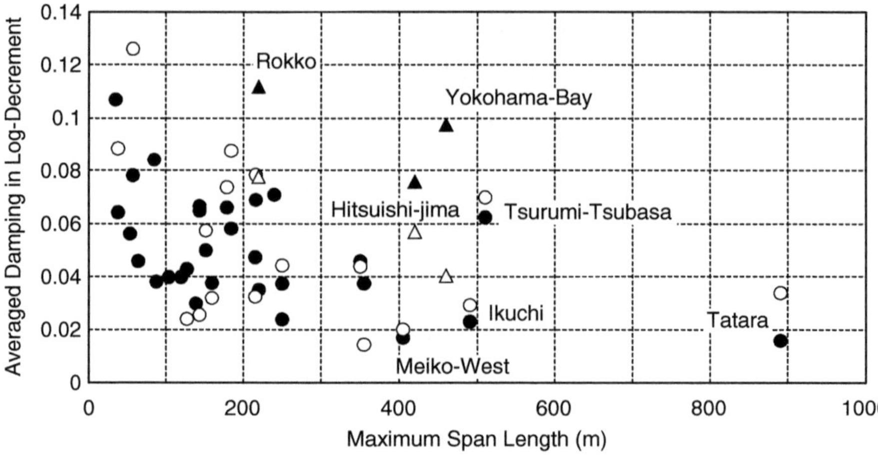

Fig. 5.4 Average damping of girder-dominant modes vs. max. span length in cable-stayed bridges: (●) vertical mode, (○) torsional mode, and (▲△) for truss girder bridge

Table 5.4 Structural damping of pylon (single column)-dominant mode in cable-stayed bridges

Bridge name	Aratsu br.	Katsushika-Harp br.	Sakitama br.	Ban-naguro br.
Pylon height (m)	60	65	80	45
1st Out-of-plane (logarithmic decrement)	0.009, 0.013	0.019	0.05	0.0169
Connection of blocks	Metal touch	Welding	Welding	Welding

5.3.2 Structural Damping for Pylon-Dominant Modes

5.3.2.1 Pylon-Dominant Modes of Cable-Supported Bridge

WRDSHSB (2001) specifies the structural damping for the pylon-dominant vibration modes with the configuration of the pylon after completion, based on the design experience of wind-induced vibrations in the Akashi-Kaikyo, Tatara, and Kurushima-Kaikyo Bridges. As for pylons in suspension bridges and A-shaped pylons in cable-stayed bridges, however, field vibration tests of bridges have seldom been conducted and the structural damping of pylon-dominant modes of those bridges have not been clearly identified yet. On the contrary, the structural damping of a free-standing single column pylon for cable-stayed bridges has been measured in the field and is in most cases less than 0.02 in log-decrement, except for some limited cases, where large value of 0.05 as shown in Table 5.4 is reported.

5.3.2.2 Free Standing Pylons During Erection

Free standing pylons during erection of cable-supported bridges have been tested in field and their structural damping data are available. Those data for long-span cable-supported bridges are listed in Table 5.5 [21–23]. As shown in the table, the damping values are about 0.01 or less in log-decrement, except the case of the Higashi Kobe Bridge. The Hakucho Bridge has significantly small value of around 0.005 and this might be because the pylon was relatively slender and fully welded. The Akashi-Kaikyo Bridge and the Kurushima-Kaikyo Bridges also show small damping for their free standing pylons of less than 0.01, because of their slender, single-box column structure. It has been found in the field vibration tests, however, that the structural damping can be significantly influenced by the existence of erection machines and also by the level of vibration amplitude. Note that what are given in Table 5.5 are the minimum values for each case.

5.3.2.3 Structural Damping of Pylon-Dominant Modes Specified in Design Codes

The field measurements have indicated that the structural damping of cable-stayed bridges with single column pylons, for the case of pylon-dominant vibration, is similar to that of girder-dominant modes in all cable-supported bridges with box girders. Based on this observation, the log-decrement of 0.02 is set as the standard value for all of these vibration modes in WRDSHSB (2001). This value is also applicable to pylon-dominant modes of bridges under construction, such as pylon-cable system and/or pylon-cable system with the partly erected girder. As for the free standing pylons under construction, on the contrary, the value of 0.01 in logarithmic decrement is specified in the design code. This is based on some engineering consideration that very low damping values have been observed in a few limited cases. It should be noted that any deviation in specifying the structural damping from the reality, for the purpose of aerodynamic design of free standing pylon, has never been attested in full scale, because the pylons in reality are usually equipped with vibration control devices.

5.3.3 Structural Damping of Stay Cables and Suspension Hangers

5.3.3.1 Structural Damping of Stay Cables

Figure 5.5 depicts the relationship between the structural damping and the natural frequency of about 50 stay-cables for the lower, in-plane modes. They all resulted from the field observations [24], though these data are not necessarily comparable to each other since the method of vibration tests and the vibration amplitude for the damping evaluation are different, test by test. However, the observation can be made as follows:

(a) The modal damping of stay cables is generally low. The logarithmic decrements are found to be less than 0.005.
(b) The modal damping for the first mode is exceptionally high but it tends to decrease for higher natural frequencies.

Table 5.5 Structural damping (in log-decrement) of free-standing pylons in cable-supported bridges

Bridge name	Akashi	2nd Kuru-shima	3rd Kuru-shima	Kanmon	Hakucho	Ohnaruto	Yoko-hama bay	Higashi-Kobe	Hitsuishi-jima	Meikou Nishi	Ikuchi	Tsurumi-Tsubasa
Pylon height (m)	283	159.95	173.45	133.8	127.9	126	172	147	143	122	120	180
Structure	Braced frame	Rigid frame	Rigid frame	Braced frame	Rigid frame	Braced frame	H-shape rigid frame	H-shape frame	Deformed frame	A-shape frame	A-shape	Inverse Y-shape
1st Out-of-plane	0.0067	0.0047	0.008	0.009	0.005	0.02	0.01	0.03	0.014	0.0078	0.013	0.014
2nd Out-of-plane	0.038	0.032	0.022	–	0.029	–	–	–	–	–	–	–
1st Torsional	0.028	0.024	0.055	0.011	0.013	–	–	–	–	–	–	–

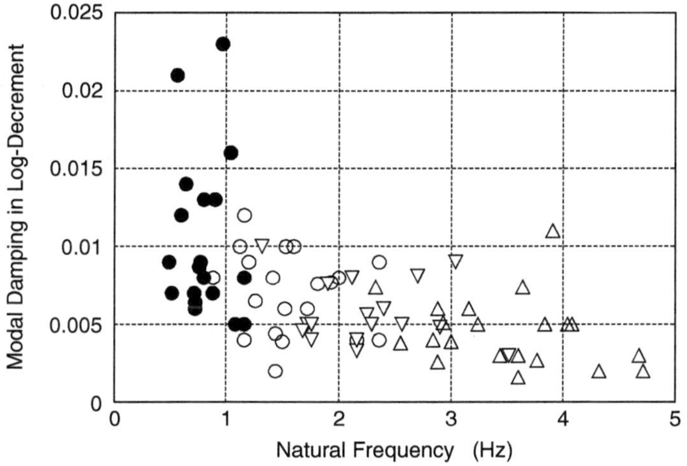

Fig. 5.5 Modal damping vs. natural frequency in stay cables of cable-stayed bridges: The 1st (●), the 2nd (○), the 3rd (▽), the 4th (△) and the 5th (□) modes

Table 5.6 Structural damping of stay cables in the Ikuchi Bridge

Stay cable		Without elastic shield			With elastic shield		
		Frequency (Hz)	Amplitude (mm)	Logarithmic decrement	Frequency (Hz)	Amplitude (mm)	Logarithmic decrement
Center span	7th Cable	0.74	16.71	0.0043	0.68	3.29	0.017
	8th Cable	0.69	20.26	0.0037	0.64	6.16	0.015
Side span	7th Cable	0.99	15.5	0.0024	0.93	1.62	0.019
	8th Cable	0.94	5.7	0.0009	0.87	5.01	0.017

Installation of dampers for preventing wind-induced cable vibrations, such as rain-wind vibration, would increase the overall damping of stay cables. In the case of the Ikuchi Bridge, the measured structural damping of stay cable was 0.003 in log-decrement for large amplitude but was increased to about 0.02 by elastically shielding the cable at its socket (Table 5.6). Similarly, in the case of the Tatara Bridge, the insertion of elastic shield material and/or high damping rubber sheet into the socket of stay cables was found to increase overall damping as shown in Table 5.7.

5.3.3.2 Structural Damping of Cable-Dominant Modes Specified in Design Codes

Referring to the above discussed characteristics of structural damping of stay cables, the standard value of structural damping for cable-dominant modes is specified as log-decrement of 0.003 in *Wind Resistant Design Standard for Honshu-Shikoku Bridges* (*2001*). This standard value is applied also for the local

Table 5.7 Structural damping (in log-decrement) of stay cables in the Tatara Bridge

Cable C38 (L = 382m)	1st Mode	2nd Mode	3rd Mode	4th Mode	5th Mode	10th Mode	20th Mode
Without control	0.011	0.005	0.0035	0.003	0.0046	0.003	0.0022
With elastic shield	–	–	–	–	0.0045	0.0046	0.0044
With high-damping rubber	–	–	–	–	0.0083	0.0068	0.0059

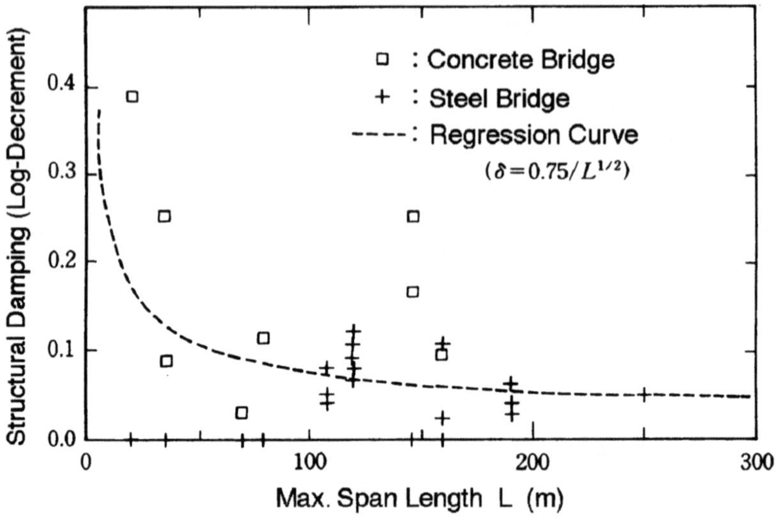

Fig. 5.6 Structural damping in log-decrement vs. max. span length in medium-span bridges [25]

vibration of hangers in suspension bridges, while the structural damping of hangers has seldom been measured in full scale.

5.4 Damping Characteristics of Girder Bridges

Measured damping data show large scatter and it is difficult to represent the structural damping by a single value for any specific bridge type. There are, however, some qualitative characteristics observed in measured data: the damping of girder bridges is larger than that of cable-supported bridges in general but becomes less with the increase of span length.

The standard values of structural damping for girder bridges are specified in *Wind Resistant Design Manual for Highway Bridges* [25] as a function of span length as shown in Fig. 5.6. The regression curve given in Fig. 5.6 and damping values obtained by shaker tests are referred to the Vibration Handbook for Civil Engineers [26].

References

1. Ito M, Yamaguchi H (1995) Full-scale measurements and structural damping of cable-supported bridges. In: Proceedings of the international bridge conference, Hong Kong, pp 557–564
2. Okauchi I, Miyata T, Tatsumi M, Kiyota R (1992) Dynamic field tests and studies on vibration characteristics of long-span suspension bridges. Struct Eng/Earthquake Eng JSCE 9(1):89s–100s
3. Narita N, Yokoyama K (1991) A summarized account of damping capacity and measures against wind action in cable-stayed bridges in Japan. In: Ito M, Fujino Y, Miyata T, Narita N (eds) Cable-stayed bridges. Elsevier, Amsterdam, pp 257–278
4. Abe M, Fujino Y (1999) Monitoring of long span suspension bridge by ambient vibration measurement. In: Proceedings of the 2nd international workshop on structural health monitoring
5. Abe M, Fujino Y, Nagayama T, Ikeda K (2003) Non-iterative identification of non-proportionally damped system from ambient vibration measurement and analysis of dynamic properties of a long-span bridge (in Japanese). J Struct Mech Earthquake Eng JSCE 745 (1–65):155–169
6. Nagayama T, Abe M, Fujino Y, Ikeda K (2005) Structural identification of a nonproportionally damped system and its application to a full-scale suspension bridge. J Struct Eng ASCE 131(10):1536–1545
7. Yamaguchi H, Fujiwara T, Yamaguchi K, Matsumoto Y (2004) Coupling of cable vibration and its damping effect in long-span cable-stayed bridge: the Tatara bridge (in Japanese). J Struct Mech Earthquake Eng JSCE 766/I-68:309–323
8. Akiyama H (1988) Field vibration tests in Honshu-Shikoku bridge authority (in Japanese). Bridge Foundation Eng 88(8):139–140
9. Yamaguchi K, Manabe Y, Sasaki N, Morishita K (1999) Field observation and vibration test of the Tatara Bridge. In: Proceedings of the IABSE conference on cable-stayed bridges, Malmo
10. Nishioka N, Toriumi R, Oka K, Sasaki N (2000) Field vibration test of the Akinada Bridge (in Japanese). In: Proceedings of the 55th annual meeting of JSCE, I-B104, pp 208–209
11. Yamaguchi H, Takahashi M (1996) Damping analysis of cable-stayed bridges. In: Proceedings of the International conference on urban engineering in Asian cities in the 21st century, Bangkok, pp 343–348
12. Yamaguchi H, Ito M (1997) Mode-dependence of structural damping in cable-stayed bridges. J Wind Eng Ind Aerod 72:289–300
13. Fujiwara T, Tamakoshi T, Ueda T, Nanjo M, Kobayashi Y (1993) Characteristics of vibration of complex multi-cable stayed bridge (in Japanese). J Struct Eng JSCE 39A:831–839
14. Yamaguchi H, Takano H, Ogasawara M, Shimosato T, Kato M, Okada J (1997) Identification of dynamic characteristics of the Tsurumi Tsubasa bridge by field vibration tests. Struct Eng/Earthquake Eng JSCE 14(2):215s–228s
15. Yamaguchi H, Takano H, Ogasawara M, Shimosato T, Kato M, Kato H (1997) Energy-based damping evaluation of cable-stayed bridges and application to Tsurumi Tsubasa bridge. Struct Eng/Earthquake Eng JSCE 14(2):201s–213s
16. Yamaguchi H, Nishimura T, Tsutsumi K, Yamaguchi K (2001) Damping effect of coupled cable vibration in a cable-stayed bridge. In: Proceedings of the 4th international symposium on cable dynamics, Montreal, pp 153–160
17. Hagiwara M, Odagiri N, Takeuchi K, Ogaki K, Hirose K (1994) Field vibration test of the Rainbow Bridge (in Japanese). In: Proceedings of the 49th annual meeting of JSCE, I-585, pp 1168–1169
18. Honshu-Shikoku Bridge Authority and Bridge & Offshore Engineering Association (1985) Construction report on the Innoshima Bridge (in Japanese). 299 pp
19. Wada K, Koyama J, Takano H, Tsumura N, Hayashi H (1992) Vibration test of the Yokohama bay bridge (in Japanese). Bridge Foundation Eng 92(2):15–18

20. Okauchi I, Miyata T, Tatsumi M, Sasaki N (1997) Field vibration test of a long-span cable-stayed bridge using large exciters. Struct Eng/Earthquake Eng JSCE 14(1):83s–93s
21. Takahashi M, Nishimoto S, Hashimoto K, Tsumura N, Koike Y (1994) Active vibration control of Hakucho bridge tower under construction (in Japanese). J Struct Eng JSCE 40A:935–942
22. Metropolitan Expressway Public Corporation (1995) The Tsurumi Tsubasa bridge, technical report (in Japanese). MEPC, Tokyo
23. Metropolitan Expressway Public Corporation (1991) The Yokohama bay bridge, technical report (in Japanese). MEPC, Tokyo
24. Yamaguchi H (1994) Modal damping of cable structures and its theory (in Japanese). Steel Construct Eng JSSC 1(3):129–138
25. Japan Road Association (1991) Wind resistant design manual for highway bridges (in Japanese). Maruzen, Tokyo, pp 119–121
26. Japan Society of Civil Engineers (1985) Vibration handbook for civil engineers, JSCE, Tokyo (in Japanese)

Chapter 6
Wind Tunnel Tests

6.1 Introduction

The wind tunnel tests have been widely used for wind resistant design of bridges, because they can provide reliable information on the wind-induced forces and responses of bridges. In this chapter, types of wind tunnel tests for wind resistant design of bridges will be listed and their brief explanations will be given first. Then, experimental results for clarifying the accuracy of the test procedure and consideration needed in order to assure the accuracy are described. Those results form the basis of the guideline of wind tunnel tests described in Sect. 3.2. Lastly, recent developments and subjects for future studies related to wind tunnel tests are stated.

6.2 Types of Wind Tunnel Tests

There are two different types of wind tunnel tests for bridges; the first type is to obtain the bridge model responses in wind, and the second is to obtain the aerodynamic force characteristics acting on the bridge model for using them in the response analysis. The tests to obtain the responses can be further classified into section model tests, taut strip model tests, and full bridge model tests, depending on how the bridge or part of the bridge is modeled. The tests to obtain the aerodynamic forces can be further divided into two: the measurement of static three components and the unsteady forces on vibrating models. Types of the tests are summarized in Fig. 6.1.

The steps in the wind resistant design procedure of long span bridges when typical wind tunnel tests are used are shown in Fig. 6.2. The section model tests of the bridge decks are used at the step when a bridge deck cross-section is chosen so that the adverse effects of wind-induced vibrations such as vortex-induced vibration or flutter on the bridge performance would be minimized. The use of section model tests to avoid wind-induced vibrations may end up giving a conservative choice.

Y. Fujino et al., *Wind Resistant Design of Bridges in Japan*,
DOI 10.1007/978-4-431-54046-5_6, © Springer 2012

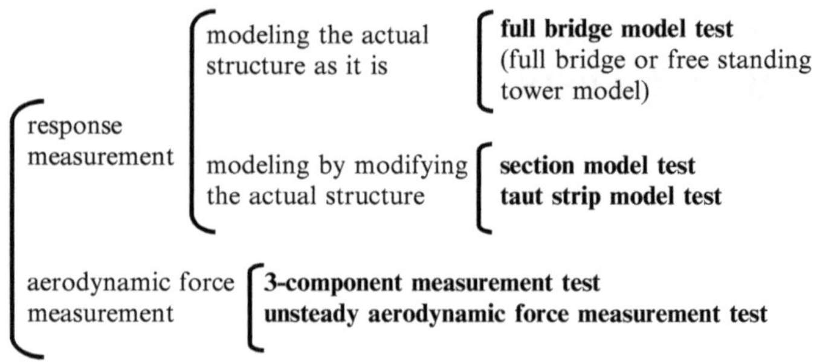

Fig. 6.1 Types of wind tunnel tests

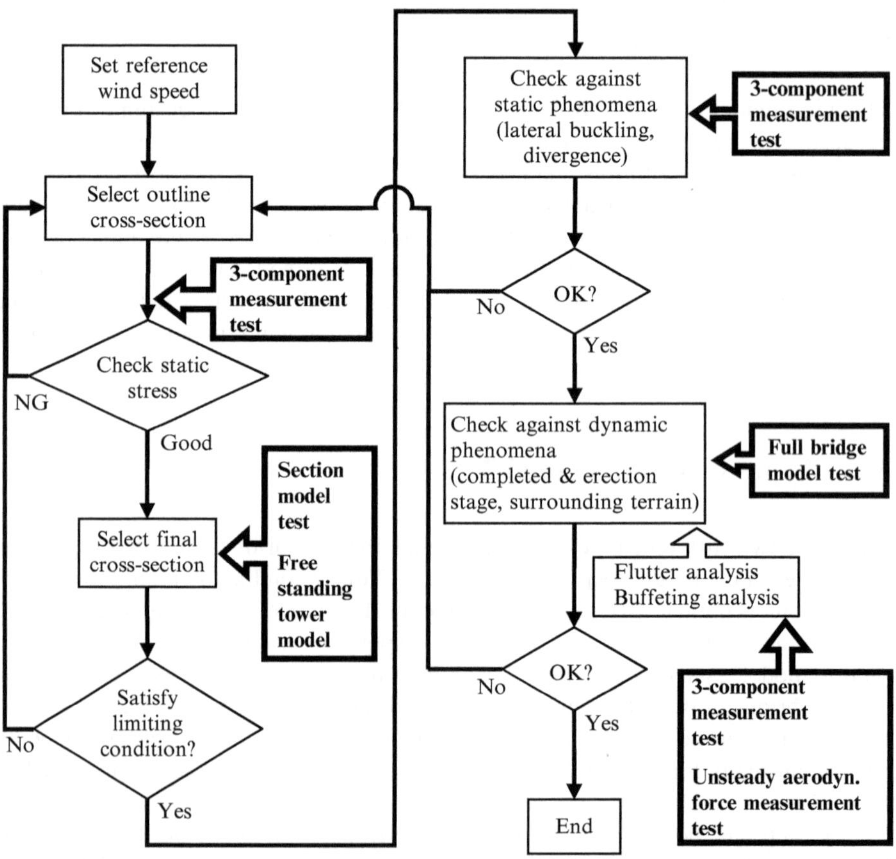

Fig. 6.2 Wind tunnel tests used in wind resistant design procedure of bridges

However, a reliable choice of the deck cross-section can be made relatively easily; the suppression of the wind-induced vibration plays dominant part of wind resistant design of long-span bridges. When the bridge span is very long, the towers would become very tall, and as a result, the wind induced response of towers need to be examined, too. Full bridge model tests may be conducted as a final confirmation of the wind resistant performance check.

The aerodynamic force coefficients obtained from the three-component tests are used in the calculation of the phenomena by static wind action, including the induced stresses, and in buffeting analysis. The unsteady aerodynamic force coefficients are used in flutter analyses that are conducted when a detailed study on the flutter characteristics is necessary.

Examples of wind flows used in wind tunnel tests are smooth flow with small turbulence, boundary layer flow that simulates natural wind, and grid generated turbulence that can easily create the required turbulence intensities. In order to clarify the detailed characteristics of some of the phenomena such as vortex-induced vibrations or flutters, the use of smooth flow is generally preferred. On the other hand, for the simulation of natural wind, the boundary layer flow would be the choice. However, it should be also noted that, for the generation of a properly simulated atmospheric boundary layer, a long test section of the wind tunnel is needed, and the scale of turbulence thus created tends to be still relatively small to conform with the required length by similitude, when the model scale is typically the order of 1:100. Grid generated turbulence can be used in a wind tunnel with a relatively short test section, but the scale of turbulence is generally small and does not satisfy the similitude requirement. Therefore, it is often used to study the effects of turbulence in a simpler manner, if possible.

6.2.1 Section Model Tests

This test uses a rigid model supported by springs that represents a part cut out from a structure such as a bridge deck or a tower. We consider here the test measuring the response of the model under wind action (Figs. 6.3 and 6.4). By modeling a part of the structure, the scale of the model can be large such as 1/50; then it is beneficial to use the section model because the details of the geometrical shape of the prototype can be reproduced relatively easily and the Reynolds number can be relatively large. Also the time and labor required for the test are much less than those for the full bridge model test that will be explained later.

In the section model tests, smooth flow is often used due to the following reasons: (1) the purpose of the experiment is to obtain the fundamental characteristics of the response of the structure, (2) it is difficult to generate turbulence that satisfied the scale requirement for the large model.

Design for the static wind loads including buffeting effects can be done with more or less satisfactory accuracy using buffeting analysis. Then the most careful attention must be paid against the dynamic phenomena such as vortex-induced

Fig. 6.3 Example of a section model test (courtesy of Yokohama National University)

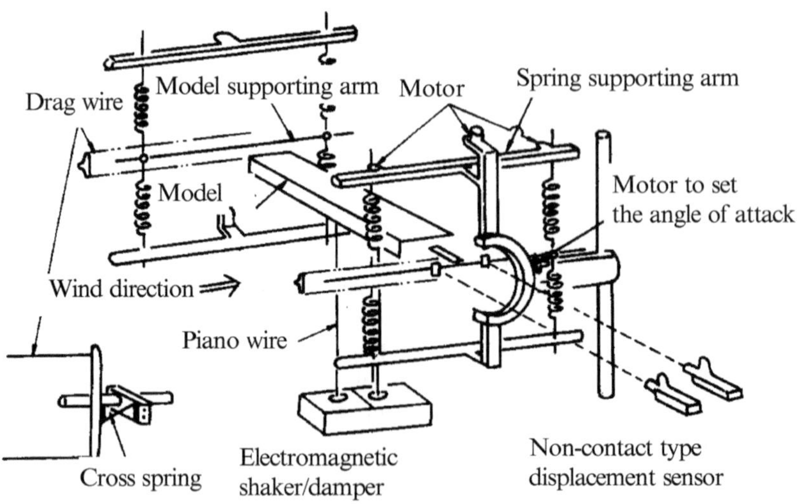

Fig. 6.4 Example of section model test setup [1]

vibration and flutter in order to assure the aerodynamic performance of bridges. These phenomena can more clearly be detected in smooth flow than in turbulent flow. Also the responses are generally larger in smooth flow with a few exceptions such as in case of vortex-induced vibration of flat hexagonal cross-section (see Sect. 8.1.1). Therefore, in order to minimize the possibility of occurrence of these phenomena, it would be necessary to clarify detailed characteristics of vortex-induced vibration and flutter in smooth flow. Note that because the actual flow acting on the bridge is turbulent, the final check of aerodynamic performance may be considered including the effects of turbulence; even if some occurrence of vortex-induced vibration is observed in smooth flow, the aerodynamic performance may be judged as satisfactory provided that the response is small enough in turbulent flow representing the natural flow at the bridge site. Even if that is the case, attention must be paid because wind with very low turbulence may act in cases such as over the sea under a special weather condition.

For the dynamic responses such as vortex-induced vibration or flutter, details of the shape such as that of handrails or edge of the deck may influence the response characteristics significantly. Section model tests are widely used for studying the occurrence possibility or suppression method of vortex-induced vibration or flutter because it can reproduce the details relatively easily due to its large scale, and also it is easy to modify the model shape when such an improvement is necessary.

Sometimes, section model tests are used for the prediction of response in turbulent flow using the grid generated turbulent flow for simplicity; in this case, because the turbulence scale similarity cannot be satisfied generally, caution is needed for the possible lack of reliability of the results.

The *Wind Tunnel Test Specification for Honshu-Shikoku Bridges* (*2001*) requires, as a standard procedure, to obtain the relationship not only between dynamic response and wind speed but also of the effective damping with them. This information is important to understand how the aerodynamic damping changes with wind speed and response amplitude, and thus the aerodynamic stability would be predicted more precisely in relation to the structural damping. The above relationship can be obtained by measuring both the damping time series of free vibration of the model from forcedly oscillated condition and the diverging time series from stable condition. In order to obtain the desired data accurately, it is necessary to have well-behaved free vibration traces. A good practice for that is to excite the model not manually but mechanically by using an electromagnetic shaker, for example.

6.2.2 Taut Strip Model Tests

Taut wires and/or tubes provide the stiffness of a taut strip model, and the model has a simple mode shape such as a half sine wave. It is mainly used to obtain the response characteristics in turbulent flow. The number of taut strip model tests conducted in Japan for practical examples is relatively small compared to the section and full bridge model tests.

Fig. 6.5 Example of free standing tower model test [2] (courtesy of Honshu-Shikoku Bridge Authority)

6.2.3 Tower Model Tests

Wind resistance of a free-standing tower at its erection stage has to be carefully studied because it is vulnerable to wind-induced vibrations (Fig. 6.5). Even at a completed cable-supported bridge, wind-induced vibration of the tower sometimes needs consideration. In order to clarify the wind resistance of a completed bridge tower in detail, the tower is modeled in a large scale and other members such as main cables are replaced with spring and/or mass system so that only their structural contributions are to be considered. For the design of the tower model after the completion of the whole bridge, it is important to perform the eigenvalue analysis of the completed full bridge, and make sure that the natural frequencies, mode shapes, and modal masses of important vibration modes of the model reproduce the required values.

For the tower of the completed Akashi-Kaikyo Bridge, the model design was studied in detail by utilizing three-dimensional FEM eigenvalue analysis; it was confirmed that the vibration modes could be reproduced with acceptable accuracy by attaching additional masses and a spring system at the top of the tower model [3]; the additional mass and stiffness had to be adjusted depending on the mode that the experimental case is to reproduce. On the other hand, the additional mass resulting from the vibrating cables and girders would be effectively reducing the dynamic response of the towers and it will be more conservative if they are ignored. Therefore, for the largest tower model with scale of 1/86 for the case of Akashi, the additional mass was not considered to avoid the complexity of the experimental procedure; but the axial force caused by supporting the main cables was considered by introducing it. As a result, by adjusting the spring stiffness and distance between springs, the appropriate additional stiffness could be set so that the out-of-plane 1st bending mode and the 1st torsional mode could be tested simultaneously.

6.2.4 Full Bridge Model Tests

Full bridge model tests (or full aeroelastic model tests) where the whole structural systems are modeled are sometimes conducted for the cases such as: long-span bridges where the wind-induced response plays dominant role in the design, and the bridges with largely varying girder cross-sections along its longitudinal axis where their wind-induced responses cannot be predicted accurately by using section models. The full bridge model test can predict the actual phenomena most correctly, provided that the actual dynamic characteristics of the prototype bridge are well reproduced; however, due to the limitation of the wind tunnel size, the full bridge model scale usually tends to be small and that often results in the shortfall such as the difficulty of the detailed model shape reproduction and too small Reynolds number. Also the full bridge model test usually requires a large wind tunnel, and the model manufacture and installation require much more labor and cost than the section model tests.

On the contrary, when the buffeting due to turbulent wind is to be reproduced, the turbulence scale similarity with the model geometrical scale can be satisfied more easily with the small-scale model. At the same time, simulation of turbulence with consistent geometrical scale with the bridge model can be achieved only by full bridge models and not by the section models. Thus the full bridge model has advantage in predicting buffeting response.

For the full bridge model test of the Akashi-Kaikyo Bridge, the priority was set to obtain accurate results with a precise model and the model scale was set to be 1/100 in order to ascertain the model precision. The total length of the model including the cable anchors became more than 40 m, and a large wind tunnel with test section width of 41 m was newly constructed. A full bridge model is often designed in such a way that a stiffness bar goes through the whole bridge length, or span, to simulate the bridge girder stiffness. The short girder modules that simulate the geometrical feature of the bridge would be attached on the stiffness bar with minimum clearances between them so that the modules give aerodynamic

Fig. 6.6 V-shape springs used for the full bridge model of the Akashi-Kaikyo Bridge (courtesy of Honshu-Shikoku Bridge Authority)

simulation but would not contribute to the overall bridge stiffness. For the model of the Akashi-Kaikyo Bridge, however, with a truss stiffening girder, it was considered undesirable to use a stiffness bar because it would be also exposed to wind and reduce the accuracy of the test results. Therefore, a stiffness bar was not used, instead v-shape springs shown in Fig. 6.6 were used to connect the modules and also to provide with the proper model stiffness. For the main cable model of the Akashi-Kaikyo Bridge, similarity of mass, tension stiffness and drag force was considered. The axial stiffness of the cable was properly simulated primarily by a thin wire extending through the whole length, and segmental small weights were attached on it to simulate the cable mass and also the wind-induced drag force on cables at the design wind speed. Only at the central part of the main span, the cables were modeled geometrically rather than considering the wind drag. This is because the section model wind tunnel tests indicated that the cables and bridge deck, which were so close to each other at this location, would cause the aerodynamic interference.

6.2.5 Three-Component Measurement Test

This test is to measure the drag and lift forces and pitching moment of a section model of the bridge stiffening girder by using a three-component balance. The aerodynamic coefficients are used to predict the static wind load and buffeting response. The force coefficients for buffeting analysis are sometimes measured in turbulent wind flow.

6.2.6 Unsteady Aerodynamic Force Measurement

For a very long-span bridge like the Akashi-Kaikyo Bridge, the characteristics of flutter that may occur become complex where many vibration modes including horizontal ones are coupled. Then the critical wind speed cannot be predicted accurately by the section model test alone that generally is a two-degree-of-freedom system with vertical and torsional motions; a three-dimensional flutter analysis has to be used together.

The unsteady aerodynamic force measurement test is to measure the unsteady aerodynamic force coefficients that are required for the flutter analysis. The coefficients have to be measured carefully because the accuracy of the measurement directly affects that of the flutter analysis. The measurement methods are classified into the forced oscillation method and the free oscillation method. The former is to give a forced sinusoidal motion to a model in wind and measure the reaction at the supports; the latter is to identify the acting forces from the free vibration response of a model in wind.

The forced oscillation method has relatively long history and there are many studies using it. The reproducibility of the result is generally good. Attention must be paid when the aerodynamic forces are measured by a balance such as a three-component balance, because the large inertial force due to accelerated motion of the model is measured at the same time. How to eliminate the inertial force determines the accuracy of the measurement. Generally, counter weight is installed and the strain gauge bridge is connected to cancel the inertial force by adjusting the location of the counter weight, then a dynamic strain amplifier can amplify the signal without the inertial force. A special apparatus that can accomplish such function is required [4]. If we remove the inertial force so that the output under still air condition becomes zero, the unsteady aerodynamic forces under still air condition are also removed. Such error may result in significant difference of unsteady aerodynamic coefficients [5]. When the bridge span length becomes very long, the unsteady aerodynamic forces caused by horizontal motion may play an important role in flutter as stated above; therefore an apparatus that can measure the unsteady aerodynamic forces under the horizontal motion has been also developed [6].

Pressure transducers are also used to measure the unsteady aerodynamic forces on the forcedly oscillated model; an advantage of this method is that the unsteady aerodynamic force can be measured without the influence of the inertial force. For a model with a complicated shape, the accuracy of this method may become worse because it is difficult to locate the pressure taps effectively for the measurement of forces. But for models with relatively simple cross-sections, the unsteady aerodynamic force can be measured accurately. For a precise measurement, it is necessary to use high precision pressure transducers that can measure small pressure with enough resolution, and to locate enough number of pressure taps on the surface of the model.

The main advantage of the free vibration method is that the unsteady aerodynamic forces can be obtained by using the same apparatus and model as for the section model test. Identification method using the extended Kalman filter [7] and a modal decomposition and reassemblage method that are applicable to a coupled vibration with multiple modes [8] have been proposed. In the modal decomposition and reassemblage method, the damping and stiffness matrices of a section model response under a wind speed are identified with extended Kalman filter by assuming that the response can be expressed as a combination of the complex eigen modes. Then a complex eigenvalue analysis is applied to the identified equation of motion, from which the complex natural frequencies and mode shapes are obtained. These steps are repeated and a series of combinations of the complex mode shapes with corresponding reduced wind speed is obtained. Then based on the equation of motion, unsteady aerodynamic force coefficients at a reduced wind speed can be obtained from the complex mode shape and eigen value at the same reduced wind speed. Furthermore, there is a study on an identification of unsteady aerodynamic forces based on rational function approximation that is applicable not only to a sinusoidal motion but also to a transient motion.

6.3 Influence of Wind Tunnel Characteristics and Blockage on the Test Results

In this section, experimental results related to the wind tunnel characteristics are described among the test results to confirm the accuracy of the wind tunnel test; differences of the same experiments' results conducted at different wind tunnels and effects of area blockage ratio (ratio of the model projection area to the area of the test section) are discussed.

6.3.1 Influence of Wind Tunnel Characteristics on the Test Results

When *Wind Tunnel Test Specification for Honshu-Shikoku Bridges* (*1980*) [1] was to be put together by examining *Standard of Wind Tunnel Test for Honshu-Shikoku Bridges* (*1973*), a concerned matter was that the test results at different wind tunnels such as the aerodynamic coefficients or wind-induced responses sometimes differed even if the same model was used. Then the effects on the test results of different wind characteristics at different wind tunnels were studied [9]. Three-component measurement and section model test results using a symmetric airfoil NACA0012 and a truss-stiffening girder at different wind tunnels are described below.

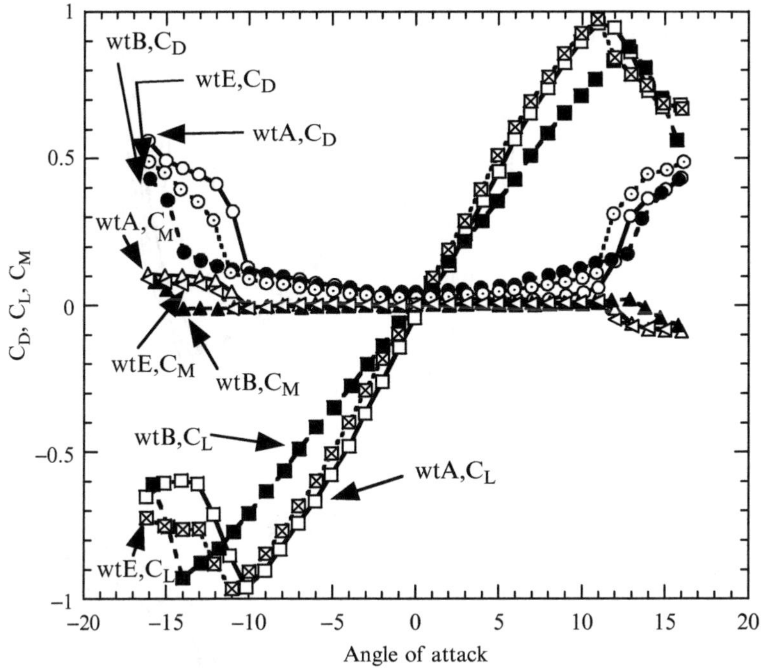

Fig. 6.7 Three-component aerodynamic coefficients of an airfoil model measured at different wind tunnels [9]

6.3.1.1 Influence on the Three-Component Aerodynamic Force Measurement

The measured aerodynamic coefficients of an airfoil model at different wind tunnels are shown in Fig. 6.7. In wind tunnel A, the lift coefficient did not pass through the origin and the wind was thought to have a slight vertical inclination when it was supposed to be horizontal. In wind tunnel B, the slope of the lift coefficient was smaller than in other wind tunnels, and possibly the two-dimensionality of the flow was not maintained possibly related to the flow turbulence or other factors. Similar tendency was observed with a truss girder model.

6.3.1.2 Influence on the Section Model Response Measurement

The airfoil and truss girder section model were elastically supported and the response was measured at the different wind tunnels with the same model mass and polar moment of inertia. Both models were designed to have their centers of gravity at the middle of their chords. Figure 6.8 shows the damping vs. wind speed for the airfoil model; the results were similar except that in wind tunnel B. The

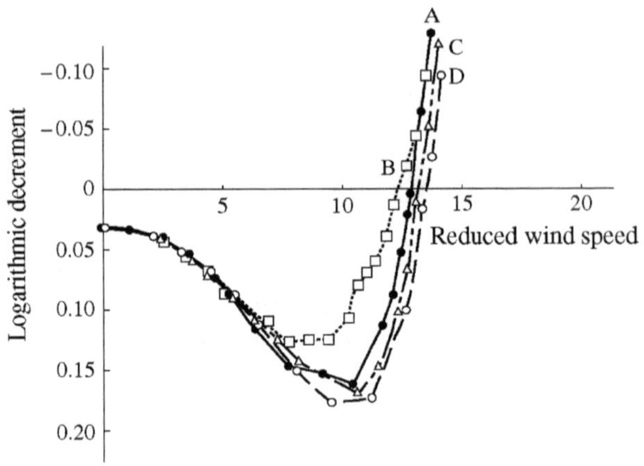

Fig. 6.8 Wind speed vs. logarithmic decrement of an airfoil model [9]

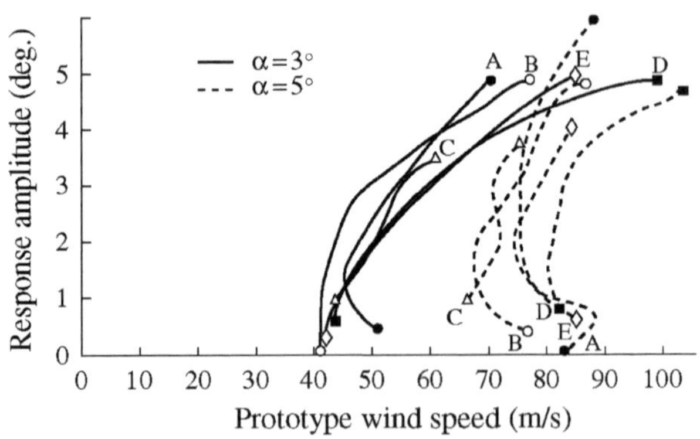

Fig. 6.9 Wind speed vs. response of a truss girder model [9]

response results of the truss girder model were more scattered (Fig. 6.9). The reason of this difference was not immediately clear but could be because of the complexity of the truss section in comparison to the wing.

6.3.1.3 Countermeasure Against the Scatter of the Results in Different Wind Tunnels

The wind tunnel B is somewhat different from other tunnels: it is an open jet type but the area blockage ratio was not very small. It was considered that the above mentioned differences in test results could have been attributed to this fact. In order to avoid such problem, a new wind tunnel to test wind-resistance of structures was constructed; the measured results conducted in this new wind tunnel were close to those in the other wind tunnels.

It was meaningful to conduct a series of tests described above when the wind tunnel procedure has to be established urgently; not only as a calibration of wind tunnel characteristics but also as to clarify the source of errors in the wind tunnel

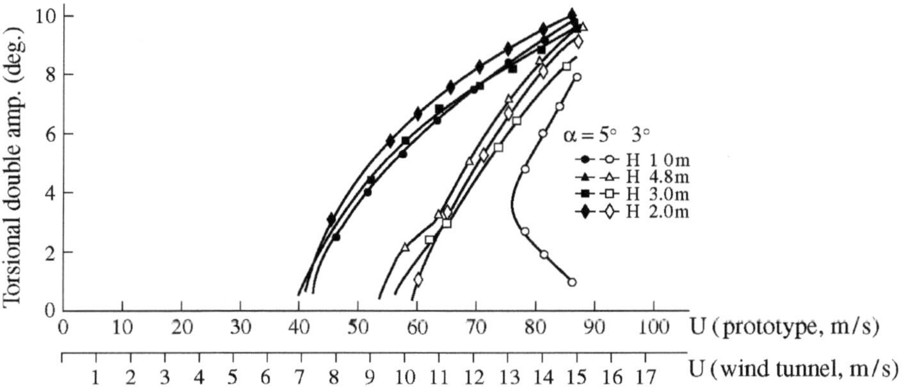

Fig. 6.10 Effects of blockage ration on the truss girder model response [9]

tests. In more recent years, the accuracy of wind tunnel tests have been much improved by restricting the area blockage ratio to less than the originally assigned value, by using the secondary end plates to avoid the effects of boundary layers developed on the wind tunnel walls, and also by setting the system's damping more precisely by the use of electromagnetic dampers. For example, the result in Fig. 6.9 is the response of the truss model with the smaller aerodynamic force, which may have caused the scatter of the test results. This scatter would be avoided if the tests are conducted now. On the other hand, further improvement of testing accuracy is still desirable and important from now on, too.

6.3.2 Influence of the Wind Tunnel Walls on the Test Results

The effects of open or closed type of test section and blockage were also studied when *Wind Tunnel Test Specification for Honshu-Shikoku Bridges* (*1980*) [1] was to be put together [9]. A section model test, a three-component measurement test, and a pressure measurement test were conducted with open (height $H = 10$ m) and closed test section (height $H = 4.8$, 3, and 2 m). The results of section model tests are shown in Fig. 6.10. At the same angle of attack, the critical wind speed of the response in the open test section was found higher. The critical wind speed was not necessarily higher in the closed test section with larger height, showing that the results with the open test section cannot be considered simply as that with a lower area blockage ratio closed test section. Figure 6.11 shows the results of three-component tests. The drag coefficient became larger with the smaller test section height, which is probably due to blockage. The lift and moment coefficients had similar tendency. On the other hand, no wind tunnel wall effects were observed in

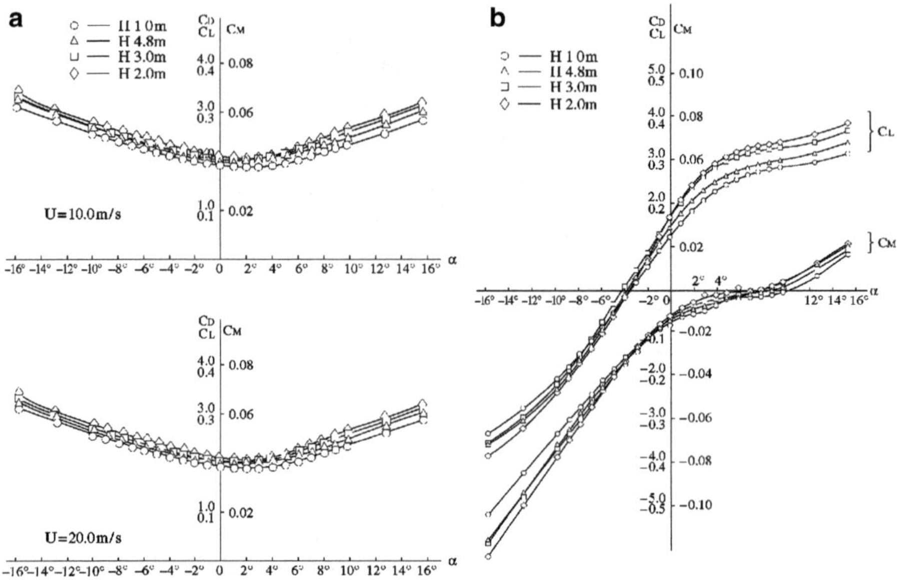

Fig. 6.11 Effects of blockage ratio on three-component aerodynamic coefficients [9]. (**a**) Drag coefficients ($U = 10$, 20m/s). (**b**) Lift and moment coefficients (U = 20m/s)

the pressure measurement. A possibility was pointed out that the model scale was 1/80 and too small to accurately reproduce the results with the section model and three-component measurement test.

Furthermore, the effects of blockage on the wind tunnel test results were examined for rectangular prisms [10]. Drag coefficient was measured for rectangular and flat hexagonal models with various blockage ratios; it was shown that the drag coefficient becomes smaller with smaller blockage ratio. A particular attention is called to the fact that this trend was observed even when the area blockage ratio was less than 5%, which is often considered to be an acceptable limit of the ratio. Then the blockage effects on vortex-induced vibration and galloping of rectangular prisms with the width-to-depth ratio of $B/D = 1$, 2, 3, and 4 were studied by changing the blockage ratio as 5%, 10%, and 15%. The blockage effects were very small for $B/D = 1$ model, but for $B/D = 2$ model, the vortex-induced vibration amplitude slightly decreased with the increase of blockage. For $B/D = 3$ model, effects of blockage are significant and the galloping which was observed with 5% blockage, did not take place when blockage was 10% or more. At the same time, the vortex-induced vibration amplitude decreases considerably with the increase of blockage. For $B/D = 4$ model, only vortex-induced vibration occured, and its amplitude also decreased considerably with the increased blockage. The effects of blockage on these wind-induced vibrations are thought to become significant when delicate flow characteristics such as flow re-attachment play important roles in the phenomena.

6.4 Reynolds Number and the Angle of Attack

6.4.1 *Influence of the Reynolds Similitude*

In the wind tunnel test for structures, similarity law of the Reynolds number is almost always ignored based on the general understanding that the effects of the Reynolds number is not significant for the flow around a body with sharp edges where the flow always separates at the same edge. However, in relation to the wind tunnel tests of the Akashi-Kaikyo Bridge towers [11], Hikami found that the amplitude of vortex-induced vibration could change considerably at the Reynolds number of 5 to 10 × 10^3, particularly when the corners of a rectangular cross-section were cut off. Alarmed by this finding, there have been many studies conducted to clarify the Reynolds number effects on wind-induced structural response [12–15].

In Fig. 6.12, the change of vortex-induced vibration amplitude vs. Reynolds number of a rectangular cross-section with corner-cuts based on a section model test is shown [11]. When the Reynolds number taking the projected height of the section as the representative length is increased from 6 × 10^3 to 10 × 10^3, the response amplitude became much larger.

The ideal way to clarify the Reynolds number effects is to compare the model test results with the measured results at the actual prototype bridge, because conducting a wind tunnel test at the actual prototype Reynolds number is generally impossible. However, it is also difficult to obtain data such as aerodynamic coefficients or wind-induced response amplitudes at the actual bridge that can directly be compared to the wind tunnel test results. Consequently, the number of such study is small; but there is a study based on a fluctuating pressure on-site measurement at the Ikara Bridge, where Strouhal number around the PC girder of the cable-stayed bridge is shown to agree approximately with wind tunnel measurements at the Reynolds number around more than 2 × 10^4 [14].

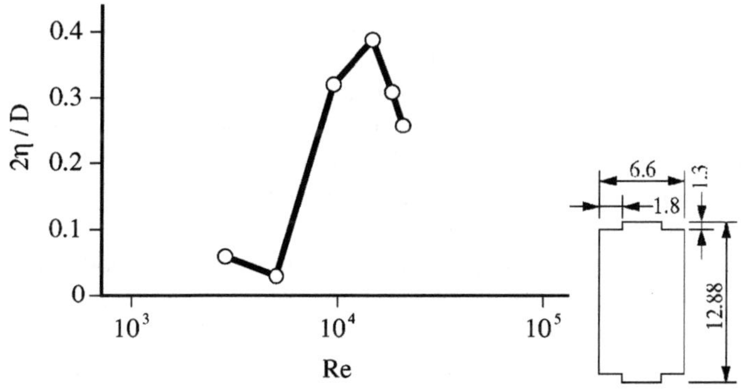

Fig. 6.12 Amplitude of vortex-induced vibration vs. *Re* of a rectangular cross-section with corner-cuts [11]

Aerodynamic force coefficients, Strouhal numbers, and unsteady aerodynamic forces were measured at high Reynolds number in the order of 10^6 by using a large wind tunnel with the test cross-section of 9 m $\times$ 9 m and a 1/10 scale large bridge girder model [15]. The section model was of a twin-box girder prepared for an ultra-long span suspension bridge. When the wind acted with the angle of attack on the model, whose gap between the box girders were covered with grating, the aerodynamic forces and Strouhal number were much influenced by the Reynolds number. However, the critical flutter speed calculated from the force coefficients measured at the higher Reynolds number became actually higher than that based on the low Reynolds number results, which ironically means that, at least for this particular case, the measurement in lower Reynolds number was more conservative.

In conclusion, the Reynolds number effects do exist on aerodynamic force coefficients and response amplitude, even when the structural cross-section has sharp edges, where the flow separation would be expected. In order to avoid errors due to the Reynolds number effects, it is thought to be advisable based on those studies that the wind tunnel test should be conducted at the Reynolds number at least more than 10^4 or so.

6.4.2 Setup of Angle of Attack in the Wind Tunnel Test

When the angle of attack of an elastically supported section model is to be set, it consists of the following factors:

1. Mean vertical inclination of the approaching wind due to the topography at the site
2. Mean torsional displacement of the suspended structure due to wind loading
3. Vertical inclination of the temporal approaching wind due to vertical fluctuation of the wind velocity in the natural wind

   For the factor (1), the angle of attack of the section model can be set according to the approaching wind characteristics at the site if they are known. But it is rare that the inclination of the wind due to topographical reasons is measured and a reliable value is obtained. The mean torsional displacement described in (2) should be considered when it is significant. For the factor (3), corresponding angle of attack cannot be set in a section model test in smooth flow. Furthermore, there is another factor due to precision of the wind tunnel test as:
4. Error in the setting of the angle of attack

   Consequently, the guideline assigns that a section model test is to be conducted for the angles of attack between -3 and $+3°$, even if the mean vertical inclination is horizontal and mean torsional displacement is small. It should be pointed out that the angle of attack of the section model has to be kept at the initial value during the test, because the torsional displacement of the section model does not represent the actual full bridge torsional displacement; when the

Fig. 6.13 Separated ground plates in span-wise direction used in a full bridge model of a Nielsen bridge in a complex terrain [11] (courtesy of Hiroshima Prefecture Public Road Corporation)

model is displaced torsionally due to the mean wind load, the setting should be adjusted so that the angle of attack is kept at the value under no wind condition.

For a full bridge model test in a simulated boundary layer flow, the factor (2) is reproduced by the structural characteristics of the model, and the factor (3) is reproduced by the flow characteristics. The factor (1) can be reproduced if the topographical model is included in the test, although limitation of the size of the wind tunnel test section sometimes does not allow its inclusion. In such a case, only the factor (4) should be considered.

In order to study the possible occurrence of vortex-induced vibration or self-excited vibrations in detail, a full bridge model test is sometimes conducted in a smooth flow. In this case, the angle of attack has to be set to reflect the factor (3). The angle of attack may be adjusted by installing airfoil cascades at the nozzle of the wind tunnel and changing their pitching angle; or by installing an inclined ground plate on the wind tunnel floor below the model. The use of the ground plate is not a primary choice since the flow inclination is given only at very low elevation, and also it is very difficult to produce negative angles against the model. Inclining the whole bridge model may be applied when it does not affect the dynamic characteristics of the model. When a bridge is located in a complex terrain, the mean vertical inclination of the approaching flow can vary along the bridge longitudinal axis; there has been a test reproducing such situation by installing ground plates separated in span-wise direction and changing their inclination one by one [16] (Fig. 6.13).

6.5 Attention Required in the Section Model Test and Studies on Its Applicability

Discussions are given on the section model test, which is an indispensable tool to understand the details of the vortex-induced vibration and self-excited vibration of bridges; similarity requirement in mass, location of the center of rotation, and results of test to clarify the section model test's applicability are discussed.

6.5.1 Similarity Requirements for Mass and Center of Rotation in Section Model Test

The wind-induced response of a full bridge system consists of various vibration modes. On the other hand, the response of a section model consists of two-degree-of-freedom, heaving and torsional, motion with uniform displacement along the model axis, because the section model is taken out from a part of the full bridge system. When the model parameters such as mass are to be determined, similarity laws have to be applied with considerations on these corresponding vibration modes; for example, the equivalent mass of the section model has to be equal to that of the full system [17]. Particularly, contributions from towers and cables to the full bridge vibration are significant in a cable-supported bridge and the average mass only per unit length of stiffening girder can be quite different from the equivalent mass. However, in *Wind Tunnel Test Specification for HSB* (*1980*), the mass of a section model is determined to be similar to the addition of the girder and cable mass, because the wind-induced vibration's shape does not necessarily agree with natural modes of vibration, and because the model with lighter mass gives safer results.

In a section model test, attention has also to be paid at where the center of rotation of the model is located. For example, a girder's torsional motion is often associated with a horizontal motion in a cable-supported bridge; then the equivalent polar moment of inertia generally becomes larger than in the case of the girder's pure torsion. In order to include such effects in a section model test, the center of rotation is determined as the mid point between the center of gravity and the shear center, in the Guideline of wind tunnel test [18]. For other types of bridges, such as the girder bridges, there is no clear understanding on where the center of rotation should be, and careful consideration is necessary.

6.5.2 Calibration of the Section Model Test by a Large Scale Bridge Model

In 1973–1975, the section model test of a very large bridge stiffening girder, with the linear scale of 1:10, was carried out in natural wind, in order to confirm the

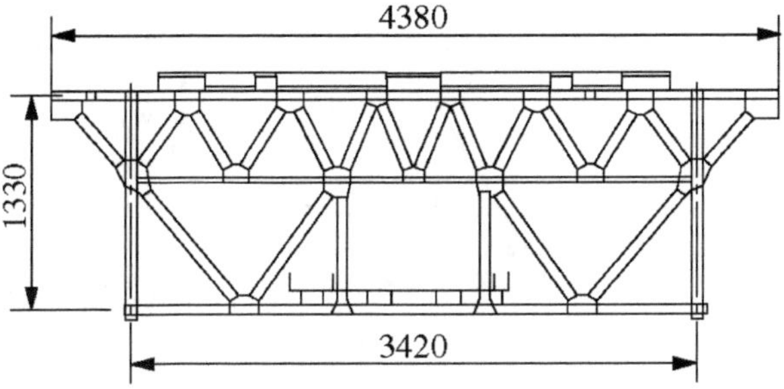

Fig. 6.14 Cross-section of the large scale bridge model (MA-1) [19]

Fig. 6.15 Large scale bridge model (courtesy of Honshu-Shikoku Bridge Authority)

Table 6.1 Drag coefficients of the large scale bridge model [19]

	MA-1	MA-2	MB-1	MB-2
Observation	2.45	2.44	2.11	2.66
Wind tunnel tests	2.27	2.31	2.13	2.67
Observation/wind tunnel tests	1.08	1.06	0.99	0.99

agreement of the section model test results with the actual bridge response [19]. The test model was a section of truss stiffening girder with length approximately of 8 m (Figs. 6.14 and 6.15). The measured drag coefficients more or less agreed with wind tunnel results (Table 6.1). The observed data were also used for the confirmation of buffeting analysis. Probable divergent amplitude type responses were observed for two cases of large scale bridge model with closed deck cross-section. The results

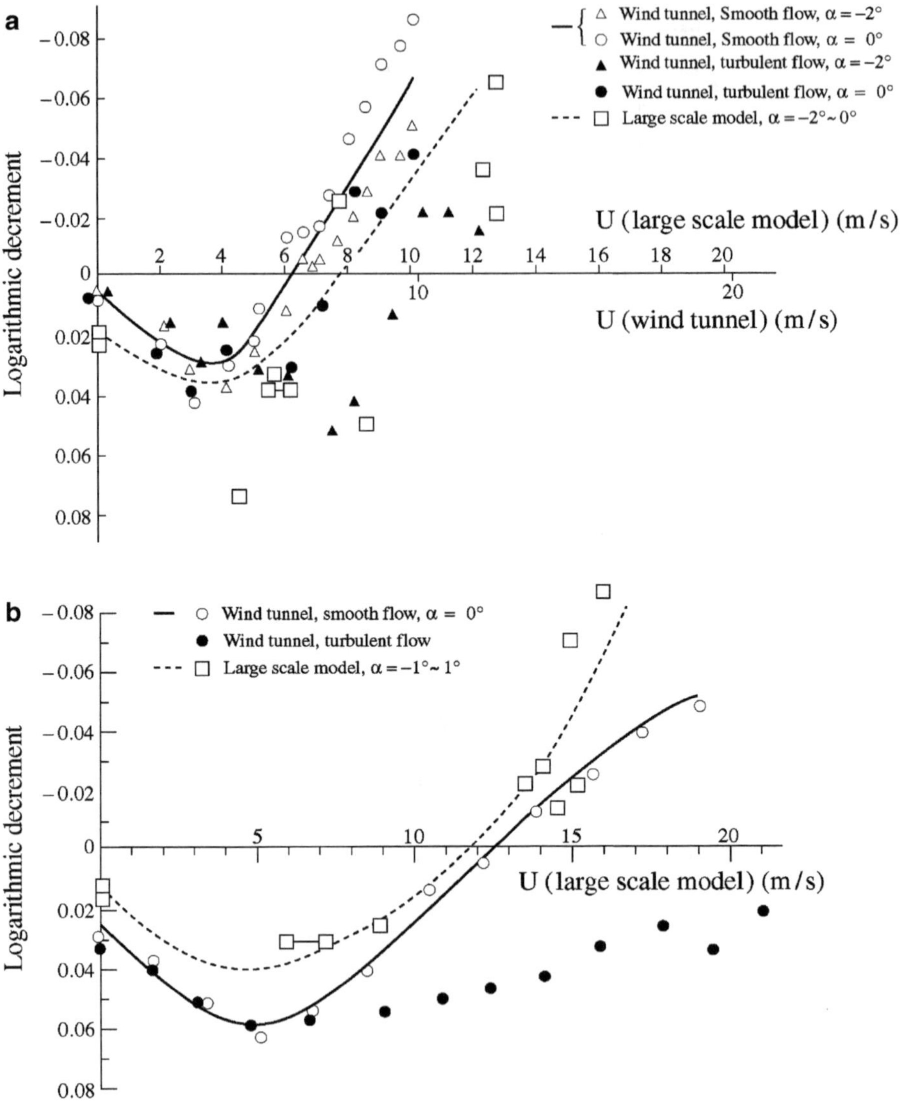

Fig. 6.16 Wind speed vs. logarithmic damping ratio [19]. (**a**) Model MA-2. (**b**) Model MB-2

are compared in Fig. 6.16 as wind speed vs. logarithmic decrement, and in Fig. 6.17 as angle of attack vs. critical wind speed for divergent type response. The wind tunnel test for comparison was conducted using models with 1/10 scale of the large scale bridge model, and both in smooth flow and the grid-generated turbulence. The observed results with the large scale bridge model generally agreed well with the wind tunnel results in smooth flow. When grating was located so that open deck

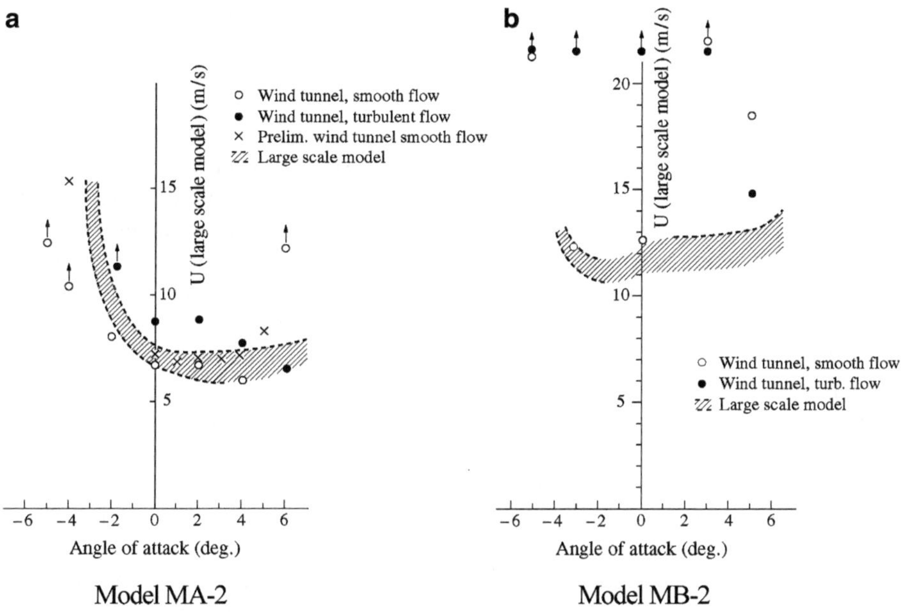

Fig. 6.17 Angle of attack vs. critical wind speed of divergent amplitude vibration [19]

cross-section was formed, no divergent type vibration was observed either in the large scale bridge model or in the wind tunnel test.

Influence of external disturbance was significant for the large scale bridge model where the data were measured outdoors; for example, the measured structural damping data tended to include some errors due to the existence of natural wind. Also, although the test section of the large scale bridge model was designed so that it produces uniform flow, there could be some effects from the bottom and top portion of the section because of the limitation of the section height. At the same time, the accuracy of the wind inclination measurement was possibly not good enough because the outdoor ultrasonic anemometer measurement was not very accurate at the early stage of the equipment's development. In Fig. 6.17, the general tendency observed on the bridge response under wind was found to be similar between the large scale bridge model and the wind tunnel test. One possible reason of quantitatively different critical wind speeds observed between them was attributed to the accuracy of the wind inclination measurement. On the other hand, it has to be pointed out that the wind tunnel results used for comparison were not so accurate as it would be expected now, because the test procedure was not fully established in those days.

By considering these situations, the agreements between the large scale bridge model and the wind tunnel results were judged to be reasonable; the wind tunnel results were considered to give the basic prediction of the actual bridge response. The cause of worse agreement of the large scale bridge model with the wind tunnel tests in grid generated turbulence compared to that in smooth flow was thought to be mainly due to mismatching of the scale of turbulence.

6.5.3 *Application of a Section Model Test for a Girder with Varying Cross-Section*

For a long-span box girder bridge, the girder cross-section often varies along the span. The possibility of predicting the wind resistance of such a varying cross-section girder by a section model test was studied [20]. The response obtained by a full bridge model test was similar to a section model test result using a uniform girder cross-section corresponding to that at L/6 span location from the span end. Actually the largest response was observed with the case among the section model test cases with the cross-sections at L/2, L/3 and L/6 span locations. The full bridge model response was different from the prediction based on the strip theory summation from the section model test results. Attention has to be paid because the predictions both by a section model test using L/3 span location that is often considered as the representative and by the strip theory can be less conservative. Also at a wind tunnel study for the Kaita Bridge, response of a full bridge model of a single-span varying cross-section girder could not be explained by a section model test result [21].

From the above cases, the applicability of a section model test is inconclusive when used to predict the wind-induced response of a girder with varying cross-section, and careful consideration is necessary.

6.6 Influence of Details of Modeling on the Test Results

The wind tunnel test results would be affected much by the details of the experimental setups, such as the end plates, dynamic characteristics of the model, and so on. In this section, influences of these details are discussed.

6.6.1 *End Plates*

Installation of the end plates at both ends of a section model is found to be useful in maintaining the two-dimensionality of the wind flow over the entire model span. Effects of the end plates were studied on the drag and Strouhal number of a plate normal to the flow, a circular cylinder, and a square cylinder [22]; by using circular end plates with the diameter of more than eight times of the model cross-section dimension in the normal direction of the approaching wind, the drag and Strouhal number become constant. But in actual cases particularly for response measurement, the smaller end plates are often preferred in order to avoid too much addition of mass. In such cases, by setting the large enough secondary end plates supported from the wind tunnel walls in addition to the small end plates attached to the model, almost the same effects as the large end plates are obtained.

Additionally, the secondary end plates are also effective in reducing the influence of the boundary layer generated on the wind tunnel walls.

6.6.2 Finish of the Model

It is considered to be very important to finish the edge of the model as precisely as possible so that the flow separation takes place exactly at the edge and also the behavior of the separated flow after that would be similar to the expected actual flow around the bridge in reality. In *Wind Tunnel Test Specification for Honshu-Shikoku Bridges (2001)*, it is specified to use hinoki (cypress japonica) as the material for the models. It is known to have a good workability and is appropriate for manufacturing models with good precision.

An error margin of the model geometry found in the *Specification* was set as the smallest value unavoidable in model fabrication process.

6.6.3 Adjustment of the Dynamic Characteristics of the Model

Another knowledge obtained from the calibration of the test using the different wind tunnels stated in Sect. 6.3.1 was the fact that the value of dynamic characteristics of the model, such as mass and stiffness, would vary as much as 10% if the adjustment method is not the same [9]. Therefore, the Specification adopts the adjustment procedure with which those parameters are identified dynamically as follows: the mass and polar moment of inertia are identified first based on the natural frequency measurement with and without known added mass. Then the stiffness is identified based on the obtained mass or polar moment of inertia and the natural frequency. The structural damping was found to depend on the vibration amplitude, and reference amplitude to set the nominal damping is assigned in the *Specification*.

6.6.4 Supporting Method of a Section Model

Attention should be paid to the fact that a section model is often supported via its end plates and the effective stiffness of the system would be reduced if the model and the end plates are poorly connected.

If only the vertical and torsional degrees-of-freedom are considered, the drag cables are often installed to avoid any stream-wise displacement under wind; the length of the drag cables has to be taken long enough so that effects on the geometrical non-linearity with a large vertical motion of the model are kept negligibly small.

Generally, the rolling motion, or a rotation with respect to the model middle axis parallel to the flow direction, is not designed to represent the actual bridge response. Therefore the rolling motion has to be suppressed to the least possible amplitude during the test and restraining wire is sometimes installed.

Related to the supporting method of a section model, a discussion arises on how the heaving vibration direction is to be modeled when the section model is set with some angle of attack. For example, the supporting springs are kept as zero angle of attack condition and only the section model is rotated to set the angle of attack; in another case, the angle of attack is set by rotating whole supporting system such as springs and drag cable. If the vertical vibration of a section model is considered, the model vibrates in the direction normal to the flow in the former case, and it vibrates in the direction normal to the model deck plane in the latter case. The issue of the directional set-up of the model for the measured results of unsteady aerodynamic forces has been discussed in a reference [23].

It is important to set the structural damping precisely to the assigned value. In a section model test, electromagnetic damper is now widely used because the damping can be adjusted easily with this equipment.

6.6.5 Seal of the Gap between the External Shape Blocks of an Elastic Model

An elastic model such as a full bridge model is often designed to attach the deck modules to a stiffness bar that simulates the structural stiffness. There should be a small clearance or a gap between adjacent modules in order not to contribute to additional stiffness and to decrease structural damping. A question is if these gaps would make any difference in the wind-induced behavior of the bridge model. A full bridge model with a box girder of varying depth along the span and an elastic model of a free-standing tower were tested in wind tunnels to investigate this issue [24]. Significant difference was observed of the model response between the models with and without sealing of these gaps. It was noted that the structural damping would be somewhat increased when the gaps were sealed. However, the main feature of the difference was that (a) the vortex-induced oscillation was observed only with the gaps sealed, and (b) the critical speed for a self-excited vibration was lower when the gaps were sealed. This result indicates that careful attention should be given regarding the sealing of gaps between deck modules, particularly when the deck height varies along the span.

6.6.6 Modeling of Gratings

In the *Specification* (*1980*), it is specified that the floor gratings can be modeled by using a flat net with the same solidity ratio. On the other hand, the shape of the grating model has come to be known to affect the section model test result when the grating plays a significant role in the aerodynamic characteristics of the model, such as the case where it covers a large opening of the deck. For example, when the grating is used for the entire deck slab, the critical wind speed for flutter of a section model with the geometrically correct shape grating is higher than the case with a flat net on the market instead of the grating [25]. In any case, general effects of different modeling of grating on bridge model response are not known; also the Reynolds number becomes very small with respect to the geometrically correct shape grating model and its effects may be significant; hence more studies are still necessary for clear understanding. Careful consideration is necessary on the modeling of grating, which may cause considerable error.

6.6.7 Effects of Snow Accretion

If a bridge is constructed in a cold snowy area, and strong wind is observed in winter, effects of the change of the shape due to snow accretion have to be taken into consideration. For the Hakucho Bridge, section model and full bridge model wind tunnel tests were also conducted with snow accretion [26, 27]. The shape of snow accretion was determined from that of a 1/2-scale bridge deck model positioned at the bridge site and the heaviest snowfall in the last 60 years. The snow accretion was modeled with urethane foam that is light and has similar surface roughness as snow.

6.6.8 Modeling of Surrounding Terrain

When a full bridge model response is measured with surrounding terrain model, large scale model is preferable for better geometrical shape reproduction of the full bridge model; then the size of the terrain model tends to become large. Effects of area blockage due to a three-dimensional body such as a terrain model are not known clearly; when the terrain model becomes large, the flow generated by the model has to be confirmed as similar to the actual case by comparing it with the one generated by another smaller scale terrain model without much blockage effects and also with on-site wind speed measurement if possible. Furthermore, the effects of surface roughness of the terrain model has to be studied because the roughness with geometrically correct size is difficult to make and even with "correct" surface roughness the terrain model may not produce appropriate flow characteristics; the performance of the model has to be checked by on-site measurement.

As it may be a special case, a wind tunnel test was conducted to study the effects of the water level of a reservoir on the wind-induced bridge response over the reservoir. In another example, the height of the sea level against a revetment was found to have significant effects on the wind-induced response of a nearby bridge. Like these cases, when the surrounding objects around the bridge may change the location or the shape, their effects have to be studied as carefully as possible.

6.7 Other Topics Related to Wind Tunnel Tests

6.7.1 Related Topics to Buffeting

6.7.1.1 Production of Two-Dimensional Fluctuating Wind Using Active Gust Generator

Active gust generator is a device that generates fluctuating wind velocity components in stream-wise and vertical direction by actively rotating back and forth the flat plate and airfoil cascades that are set horizontally (Fig. 6.18). The active gust generator has been used for fundamental studies on buffeting characteristics such as aerodynamic admittance of bridge deck cross-section [28–30]. The generated fluctuating wind velocity has the correlation of approximately unity in the horizontal direction normal to the flow, because the rotation angle of the flat plate and airfoil cascades is the same along that direction.

On the other hand, a device to generate a fluctuating velocity with a correlation less than unity in the horizontal direction, normal to the flow, has been developed [31]. Elastic flat plate cascades are used for this case and they are actuated with different phase along the span-wise direction (Fig. 6.19). In a boundary layer

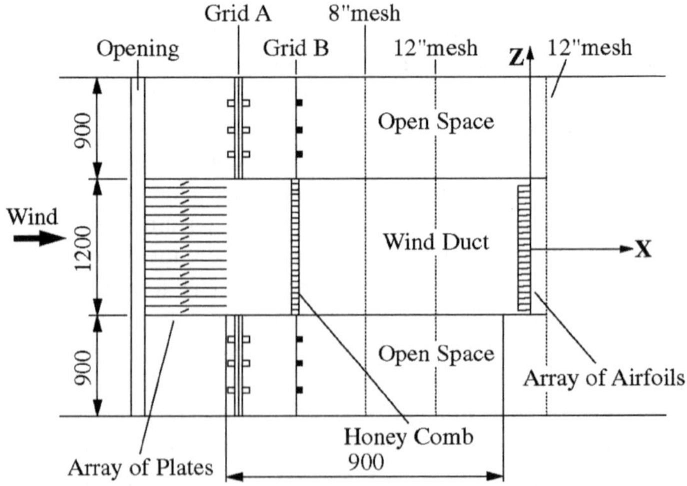

Fig. 6.18 Active gust generator [28]

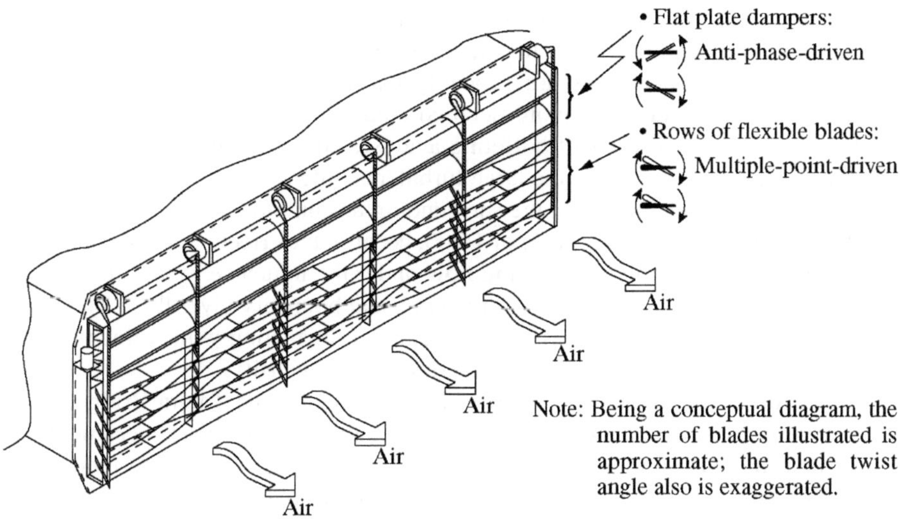

Fig. 6.19 Gust generator that can control span-wise spatial correlation [31]

turbulence generation, active spires with opening and closing motion or rotating motion, and active roughness blocks with up and down motion are studied for their effects to generate larger turbulence intensity and to control the scale of turbulence [31].

6.7.1.2 Wind Tunnel with Active Turbulence Generation

A new type of wind tunnel has been developed to generate a turbulent flow with controlled characteristics [32]. For this case, 9 columns and 11 rows of small fans, 99 in total, are installed at the intake of the wind tunnel; the small fans consisting of rapid-response servomotors and light propellers are used and the variation of the rotation speed of each fan is controlled separately. Mainly stream-wise velocity component is controlled by the rotation speed of the small fans. Active airfoil cascades that are similar to the aforementioned active gust generator, are installed simultaneously to also control the vertical velocity component.

6.7.1.3 Characteristics of Bridge Buffeting Caused by the Surrounding Terrain

The Tatara Bridge is surrounded by many small islands and a hill of nearly 500 m elevation that locates near the site. Because the surrounding terrain is considered to have significant effects on the wind flow approaching to the bridge, the Honshu-Shikoku Bridge Authority conducted a full bridge model test with 1/200 linear scale and the surrounding terrain model [33]. Particularly when the wind is from the hill, much larger buffeting than that in usual boundary layer turbulence was observed. The response particularly during the erection stage in the wake of the hill was very large. These large responses are probably caused mainly by large turbulence intensity due to the hill; but more detailed study is necessary to understand the phenomena because a series of buffeting analyses showed that the large response cannot be explained solely by the large turbulence intensity.

A full bridge model test with the surrounding terrain model was also conducted for the Megami Bridge, at the scale of 1/200, because the effects were considered to be significant [34]. In the approaching flow under the influence of the surrounding terrain, larger response was observed than in usual boundary layer flow. The cause was considered to be the increase in coherence of fluctuating wind velocity in the bridge span-wise direction as well as the turbulence intensity.

In the full bridge model test for the erection stage of the Tatara Bridge, spiky vertical responses, where the ratio of the peak to the r.m.s. response became much larger than ordinary bridge buffeting were observed, when the bridge model was located in the wake of the nearby hill. Interested in this phenomenon, the cause was studied using much simpler models [35]. On the other hand, such a spiky response was not observed for the Megami Bridge.

6.7.2 *Flow Visualization and CFD*

Flow visualization test is useful in understanding the phenomena that cause wind-induced response, of bridges. Consideration is necessary when the phenomenon's dependence on the Reynolds number is possibly significant, because the wind speed for the flow visualization test has usually to be set low in order to have clearer image.

Recent progress of computational fluid dynamics (CFD) is remarkable due to rapid improvement of computer's capability and development of computation methods. There are many studies that apply CFD to bridge deck cross-sections and that try to improve the method for bridge application [36–38]. However, it requires further improvement because the results obtained so far are not accurate enough for the quantitative prediction of wind-induced response of an actual bridge with complex geometrical shape. On the other hand, CFD has a benefit that can obtain complete information about the flow field once the computation is performed; much more information can be obtained than velocity measurements in a wind tunnel test and flow visualization test. By utilizing such characteristics together with wind tunnel test results, we may be able to have much clearer understanding of the wind-induced response of bridges. At the same time, parameters tend to be varied more easily in CFD than in a wind tunnel test. The general aerodynamic characteristics obtained from CFD are now used at the first stage of bridge deck shape selection.

References

1. Honshu-Shikoku Bridge Authority (1980) Wind tunnel test specification for Honshu-Shikoku bridges (in Japanese). HSBA, Kobe
2. Honda A, Tatsumi M, Hata K, Ohnishi E (1995) Aerodynamic characteristics of the tower of the Akashi-Kaikyo Bridge. In: Proceedings of the 9th international conference on wind engineering, New Delhi, India, pp 803–814
3. Okano S, Kurino S, Morishita T (1990) Wind-proof design of Akashi-Kaikyo Bridge (selection of pillar section) (in Japanese). Honshi Technical Report, BOEA 14 (54): 11–24
4. Okubo T, Narita N, Yokoyama K (1975) Some approaches for improving wind stability of cable-stayed girder bridges. In: Eaton KJ (ed) Proceedings of the 4th international conference on wind effects on buildings and structures, Cambridge University Press, Heathrow, UK, pp 241–249
5. Kubo Y, Ito M, Miyata T (1976) Nonlinear analysis of aerodynamic response of suspension bridges in wind (in Japanese). In: Proceedings of the Japan society of civil engineers (252):35–46
6. Tanaka H, Hatanaka A (2000) Prediction method of aerodynamic admittance function using flutter derivatives (in Japanese). J Wind Eng JAWE (83):141–160
7. Iwamoto M, Fujino Y (1995) Identification of flutter derivatives of bridge deck from free vibration data. J Wind Eng Ind Aerod 54–55:55–63
8. Yamada H, Miyata T (1997) Introduction of a modal decomposition and reassemblage method for the multi-dimensional unsteady aerodynamic force measurement. J Wind Eng Ind Aerod 69–71:769–771
9. Task Committee for Wind Resistant Design of Honshu-Shikoku Bridges (1979) Report of studies on wind resistance of Honshu-Shikoku Bridges (in Japanese). HSBA, Kobe
10. Takeda K, Kato M (1992) Wind tunnel blockage effects on drag coefficient and wind-induced vibration. J Wind Eng Ind Aerod 41–44(2):897–908
11. Kitagawa M, Akiyama H, Toriumi R (1996) Main tower of the Akashi-Kaikyo Bridge (in Japanese). In: Proceedings of the briefing session of research committee on wind tunnel experiment similarity law (2nd term), committee on structural engineering JSCE, pp 60–61
12. Report of Research Committee on Wind Tunnel Experiment Similarity Law (1994) Committee on Structural Engineering JSCE (in Japanese)
13. Okajima A, Ueno H, Abe A (1991) Influence of Reynolds number on flow and aerodynamic characteristics of bluff bodies with rectangular section of cut corners (in Japanese). J Wind Eng JAWE (49):1–13
14. Kubo Y, Nogami C, Yamaguchi E, Kato K, Niihara Y, Hayashida K (1999) Study on Reynolds number effect of a cable-stayed bridge girder. In: Larsen A, Larose GL, Livesey FM (eds) Proceedings of the 10th international conference on wind engineering: wind engineering into the 21st century. Balkema, Copenhagen, Denmark, pp 935–940
15. Matsuda K, Cooper KR, Tanaka H, Tokushige M, Iwasaki T (1999) An investigation of Reynolds number effects on the steady and unsteady aerodynamic forces on a 1:10 scale bridge deck section model. In: Larsen A, Larose GL, Livesey FM (eds) Proceedings of the 10th international conference on wind engineering: wind engineering into the 21st century. Balkema, Copenhagen, Denmark, pp 971–978
16. Uejima H, Matsuda K, Takamori S, Ogihara K, Wakisaka H, Terada H, Higuchi T (2002) Wind tunnel test for a Nielsen bridge using inclined plates. In: Proceedings of the 2nd international symposium on advances in wind and structures, Pusan, Korea
17. Yamada H, Tanaka H (1987) On predicting the performance under wind of full bridges from section model wind tunnel results. J Wind Eng Ind Aero 26(3):289–306
18. Yamada H (1995) Approach to wind resistant engineering (in Japanese), Kensetsu-Tosho, Tokyo

19. Okauchi I, Tajima J, Akiyama H (1979) Response of the large scale bridge model to natural wind. In: Wind engineering, Proceedings of the 5th international conference. Pergamon, Fort Collins, Colorado, pp 841–852
20. Saito T, Honda A (1990) On aerodynamic stability of long-span box girder bridge and anti-vibration design consideration (in Japanese). J Struct Eng JSCE 36A:889–894
21. Higaki T, Yoshihara M, Hiragi S (1989) Design and construction of the Kaita Bridge (super structure) (in Japanese). Bridge Engrg 25(12):2–18
22. Kubo Y, Miyazaki M, Kato K (1989) Effects of end plates and blockage of structural members on drag forces. J Wind Eng Ind Aero 32(3):329–342
23. Yamada H, Katsuchi H, Suzuki T (2000) Experiment on unsteady aerodynamic force and flutter response estimation under different experimental settings (in Japanese). J Struct Eng JSCE 46A:1023–1028
24. Saito T, Fujimoto N, Honda A, Watanabe K (1990) Wind tunnel method using a 3-dimensional aeroelasitic model–effects of gap between the external shape blocks (in Japanese). In: Proceedings of the 45th annual meeting of JSCE I-431:892–893
25. Ito S, Osako S, Miyazaki M, Saito Y (2000) Experimental study on the aerodynamic characteristics of a grating deck model of ultra-long span bridge (in Japanese). In: Proceedings of the 55th annual meeting of JSCE, I-B22
26. Wada T, Nosaka R, Sato K, Yoneda Y (1989) Aerodynamic stability of the stiffening girder and main tower of the Hakucho Bridge (in Japanese). In: Proceedings of the 44th annual meeting of JSCE I-401:854–855
27. Saito Y, Takahashi M, Yoneda Y (1991) Aerodynamic stability of the Hakucho Bridge (in Japanese). In: Proceedings of the 46th annual meeting of JSCE 259:556–557
28. Kobayashi H, Hatanaka A, Ueda T (1994) Active simulation of time histories of strong wind gust in a wind tunnel. J Wind Eng Ind Aero 53(3):315–330
29. Kawatani M, Kim H (1992) Evaluation of aerodynamic admittance for buffeting analysis. J Wind Eng Ind Aero 41(1–3):613–624
30. Hatanaka H, Tanaka H (2002) New estimation method of aerodynamic admittance function. J Wind Eng Ind Aero 90(12–15):2073–2086
31. Fujimoto N, Saito T, Honda A (1997) Natural wind active simulation technique for shapening wind tunnel on bridges. In: Volume of abstracts, 4th Asia-Pacific symposium on wind engineering, Gold Coast, pp 227–229
32. Nishi A, Kikugawa H, Matsuda Y, Tashiro D (1999) Active control of turbulence for an atmospheric boundary layer model in a wind tunnel. J Wind Eng Ind Aero 83:409–419
33. Akiyama H (1999) Wind resistant design of the Tatara Bridge. In: Fujisawa N, Yamada H, Miyata T (eds) Long-span bridges and aerodynamics. Springer, Tokyo, pp 266–278
34. Honda A, Fukahori S, Imagane S, Ishioka F (1999) Aerodynamic design of cable-stayed bridge surrounded by undulated topography. In: Larsen A, Larose GL, Livesey FM (eds) Proceedings of the 10th international conference on wind engineering: wind engineering into the 21st century. Balkema, Copenhagen, Denmark, pp 913–918
35. Roy UK, Kimura K, Fujino Y (1999) Vertical response characteristics of a flat plate cantilever in the wake of a hill. In: Larsen A, Larose GL, Livesey FM (eds) Proceedings of the 10th international conference on wind engineering: wind engineering into the 21st century. Balkema, Copenhagen, Denmark, pp 1035–1040
36. Kuroda S (1998) An application of Navier–Stokes simulation in bridge aerodynamics. In: Larsen A, Esdahl S (eds) Bridge aerodynamics. Balkema, Copenhagen, Denmark, pp 337–347
37. Shimada K, Ishihara T (1999) Prediction of aeroelastic vibration of rectangular cylinders by K-ε model. J Aero Eng ASCE 112:95–107
38. Structures Research Group of Public Research Works Institute et al (2002) Report of cooperative research on wind resistant design of ultra-long span bridges considering economical efficiency— part 2 (in Japanese). Cooperative Research Report No 279, Public Works Research Institute

Chapter 7
Numerical Analysis and Its Verification in Wind Resistant Design

7.1 Introduction

As the span length of a bridge grows, wind effects on the structure would become more serious, even coming to govern the structural design. Ever since the collapse of the Tacoma Narrows Bridge signified the importance of dynamic wind effects, wind tunnel tests have been often carried out to verify aerodynamic stability of long-span bridges. However, since the wind-induced phenomena of long-span bridges are interaction between wind and the structure, the exact modeling of phenomena is not without difficulty even today. Analytical efforts in the past to predict the wind-induced phenomena of long-span bridges have never been up to our present satisfaction.

The maximum dynamic wind load assumed on a long-span bridge has been evaluated based on Davenport's buffeting analysis [1–4] in Japan. This analysis is supported by the development of random vibration theory and the quasi-steady approach for buffeting forces. Therefore, the accuracy of the buffeting analysis affects structural design of a long-span bridge of which some structural members are often decided by the wind load. More precise analytical method and verification of the accuracy of analysis are considered to be an important task. For this purpose, monitoring of the Honshu-Shikoku Bridges has been continued since their completion and comparison of analytical prediction with the collected field data has been carried out. In addition, a wind tunnel test [5–10] with a 40 m long full aeroelastic model of the Akashi-Kaikyo Bridge has produced a number of findings and insights into buffeting mechanism of a long-span suspension bridge.

As for flutter, wind tunnel investigation [2–4, 11] started immediately after the Tacoma Narrows Bridge incident. Many wind tunnel tests have been also carried out in Japan for long-span bridge projects such as Honshu-Shikoku bridges, including both the section model tests and full aeroelastic model tests. In particular, the full-model wind tunnel tests of the Akashi-Kaikyo Bridge [5–10] were most extensive. Flutter prediction by numerical analysis also has a long history. Scanlan applied an experimental approach to define flutter derivatives and established a

Y. Fujino et al., *Wind Resistant Design of Bridges in Japan*,
DOI 10.1007/978-4-431-54046-5_7, © Springer 2012

standard method of flutter analysis by using them [12–14]. Many practical researches of flutter analysis based on the Scanlan approach have been carried out in Japan. Of particular interest is that the results of flutter analysis for the full bridge wind tunnel model of the Akashi-Kaikyo Bridge compared with the observed test results. Thus the analytical method was experimentally verified [8–10]. The flutter analysis was used as a tool that converted the results of the wind tunnel tests to the full-scale prediction. As a result, a methodology of flutter analysis was established in Japan.

When the bridge span becomes very long, it becomes more important to consider the three-dimensionality of the bridge behavior in detail. The analytical approach, hence, becomes more useful since the change in structural dimensions can be easily handled in this approach without going through time-consuming and expensive alterations of physical models. In comparison to buffeting analysis, flutter analysis has been more recently recognized as a design tool. For example, there are some long-span bridge projects under planning in Japan [15, 16]. All of them would involve even larger bridges than the Akashi-Kaikyo Bridge. In order to reach the final design proposals from a large number of alternatives, exhaustive investigations by applying flutter analyses have been carried out.

Vortex-induced vibration is also a concern of bridge designers. However, the usual practice in Japan is to investigate it by conducting wind tunnel tests, though, there have been some formulae for the prediction of maximum-amplitude induced by vortex excitation proposed in some wind resistant design codes [17, 18]. Research on the analytical estimation of vortex-induced vibration has also been extensively made.

This chapter describes a general theory of flutter and buffeting analysis together with their applications to practical design of long-span bridges in Japan. In addition, some important facts found in verification works are presented. Finally, the analytical treatment of vortex-induced vibration is introduced.

7.2 Flutter Analysis

Equation of motion for a three-dimensional finite element model of a long-span bridge under wind loading can be written as follows:

$$\mathbf{M}\ddot{\mathbf{u}} + \mathbf{C}\dot{\mathbf{u}} + \mathbf{K}\mathbf{u} = \mathbf{F_v}\dot{\mathbf{u}} + \mathbf{F_d}\mathbf{u} + \mathbf{F_b} \tag{7.1}$$

where $\mathbf{M}$, $\mathbf{C}$ and $\mathbf{K}$ are the mass, damping and stiffness matrices, respectively, and $\mathbf{u}$ is the displacement vector. $\mathbf{F_v}$ and $\mathbf{F_d}$ are the motion-dependent unsteady aerodynamic force matrices associated with velocity and displacement, respectively. $\mathbf{F_b}$ is the buffeting force vector. $(\dot{\bullet})$ represents differentiation with respect to time.

The unsteady aerodynamic force terms of (7.1) are experimentally obtained by wind tunnel tests using a sectional bridge deck model. Aerodynamic lift (L) and pitching moment (M) associated with the vertical (y) and torsional (θ)

degrees-of-freedom are essentially important in this analysis and usually experimentally obtained. The aerodynamic drag terms (D) and other force terms related to the lateral (z) degree-of-freedom are sometimes, but not always, included in the force matrices. The measured aerodynamic forces are represented by the non-dimensional flutter derivatives as shown in (7.2). Multiplying by 2π, these derivatives are converted to the conventional flutter derivative notations defined by Scanlan et al. [13, 14].

$$L = \pi\rho B^3\omega^2\left[L_{yR}\frac{y}{B} + L_{yI}\frac{\dot{y}}{B\omega} + L_{ZR}\frac{z}{B} + L_{ZI}\frac{\dot{z}}{B\omega} + L_{\theta R}\theta + L_{\theta I}\frac{\dot{\theta}}{\omega}\right] \tag{7.2a}$$

$$D = \pi\rho B^3\omega^2\left[D_{yR}\frac{y}{B} + D_{yI}\frac{\dot{y}}{B\omega} + D_{ZR}\frac{z}{B} + D_{ZI}\frac{\dot{z}}{B\omega} + D_{\theta R}\theta + D_{\theta I}\frac{\dot{\theta}}{\omega}\right] \tag{7.2b}$$

$$M = \pi\rho B^4\omega^2\left[M_{yR}\frac{y}{B} + M_{yI}\frac{\dot{y}}{B\omega} + M_{ZR}\frac{z}{B} + M_{ZI}\frac{\dot{z}}{B\omega} + M_{\theta R}\theta + M_{\theta I}\frac{\dot{\theta}}{\omega}\right] \tag{7.2c}$$

In (7.2), ρ is the air density, B is the deck width, K ($= \omega B/U$) is the reduced frequency, ω is the circular frequency and U is the mean wind speed. L_{yR} L_{yI}, ..., $M_{\theta I}$ are the unsteady aerodynamic force coefficients or the flutter derivatives of the given deck cross-section and are functions of K.

Since the unsteady aerodynamic forces which are in the right-hand side of (7.1) are functions of the velocity and displacement, by transferring them to the left-hand side, (7.1) can be rewritten as follows:

$$\mathbf{M}\ddot{\mathbf{u}} + \mathbf{C}'\dot{\mathbf{u}} + \mathbf{K}'\mathbf{u} = \mathbf{F_b} \tag{7.3}$$

where $\mathbf{C}' = \mathbf{C} - \mathbf{F_v}$, $\mathbf{K}' = \mathbf{K} - \mathbf{F_d}$.

The modal analysis is now applied to (7.3). Since the right-hand side of (7.3) is the buffeting force vector, which is a non-homogeneous part of the aerodynamic forces and does not make the system unstable, it is commonly omitted for performing the flutter analysis.

Expressing the displacement vector by the multiplication of a mode shape matrix $\mathbf{X}$ and the generalized coordinate vector $\mathbf{\Phi}$

$$\mathbf{u} = \mathbf{X\Phi} \tag{7.4}$$

Equation (7.3) can be transformed as

$$\tilde{\mathbf{M}}\ddot{\mathbf{\Phi}} + \tilde{\mathbf{C}}\dot{\mathbf{\Phi}} + \tilde{\mathbf{K}}\mathbf{\Phi} = \mathbf{0} \tag{7.5}$$

where $\tilde{\mathbf{M}} = \mathbf{X}^T\mathbf{MX}$, $\tilde{\mathbf{C}} = \mathbf{X}^T\mathbf{C}'\mathbf{X}$, $\tilde{\mathbf{K}} = \mathbf{X}^T\mathbf{K}'\mathbf{X}$, which are the generalized mass, damping and stiffness matrices, respectively.

Assuming a sinusoidal motion for the generalized coordinate vector $\mathbf{\Phi}$ as

$$\mathbf{\Phi}(t) = \mathbf{\Phi}\exp(\lambda t), \;\; \lambda = \lambda_R + i\lambda_I \tag{7.6}$$

a flutter condition can be deduced, from (7.5), as follows:

$$\det\left(\lambda^2 \tilde{\mathbf{M}} + \lambda\tilde{\mathbf{C}} + \tilde{\mathbf{K}}\right) = 0 \tag{7.7}$$

Equation (7.7) can be solved as a complex eigenvalue problem in which the circular frequency ω and critical damping ratio ξ can be calculated from the eigenvalue $\lambda(= \lambda_R + i\lambda_I)$ and eigenvector $\mathbf{\Phi}(= \mathbf{\Phi}_R + i\mathbf{\Phi}_I)$:

$$\omega = \sqrt{\lambda_R^2 + \lambda_I^2} \tag{7.8}$$

$$\xi = \frac{\lambda_R}{\sqrt{\lambda_R^2 + \lambda_I^2}} \tag{7.9}$$

The wind speed U at which the damping ratio, or the real part of the eigenvalue λ_R, becomes zero is the flutter critical wind speed. The frequency of flutter can be obtained by the imaginary part, λ_I, at that wind speed. Since (7.7) is a function of the reduced frequency K, there are two ways to obtain the flutter speed and corresponding frequency. One is that the circular frequency ω and wind speed U are assumed at first and a calculation is repeated until assumed ω and U converge to the roots by (7.8) and (7.9). The other is that U is calculated by ω obtained, assuming K.

The procedure described above has an advantage of shorter calculation time because of the usage of modal analysis. However, a key to the success of this analysis is the usage of adequate vibration modes. Therefore, unlike the modal analysis method, a direct flutter analysis method [19–21], in which an eigenvalue problem is directly applied to (7.5), was developed. Earlier studies have shown that the outcome of the flutter analysis by the modal method with adequate number of modes coincide with those of the direct flutter analysis.

$\mathbf{F_v}$ and $\mathbf{F_d}$ in (7.1) are the motion-dependent unsteady aerodynamic force matrices associated with velocity and displacement, respectively; and are often called flutter derivatives. Since flutter derivatives are functions of the reduced wind speed, iterative processes are required to obtain the flutter critical wind speed. Rational function approximation (RFA) is available to describe the flutter derivatives. RFA has certain advantages that it allows state-space time-domain analysis, and therefore does not require iteration. RFA shows advantages in dealing with active flutter control (for example [22–24]) and nonlinearity of bridge structures.

7.2.1 Flutter Analysis for the Akashi-Kaikyo Bridge

As already described, the flutter analysis was utilized as a tool that converted the full-model wind tunnel test results of the Akashi-Kaikyo Bridge to the full-scale

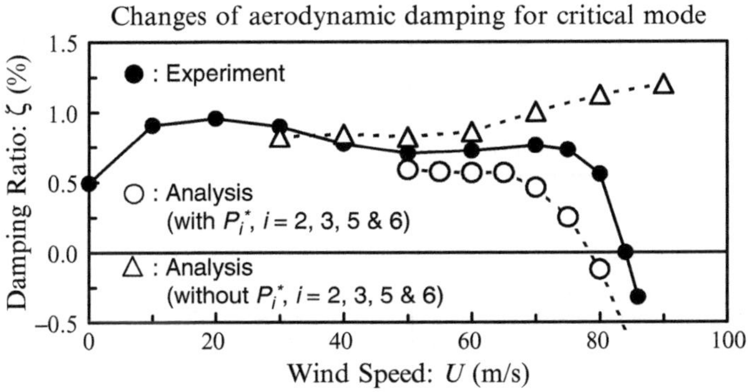

Fig. 7.1 Comparison between flutter analysis and full-model wind tunnel test of Akashi-Kaikyo Bridge

bridge behavior. It gave great insights into the coupled flutter of a long-span suspension bridge. In addition, by comparing the results of the wind tunnel tests with flutter analysis, some important facts were pointed out concerning the methodology of flutter analysis. The standard understanding about the methodology of flutter analysis was thus established.

Figure 7.1 shows the comparison between the results of flutter analysis and the full-model wind tunnel test of the Akashi-Kaikyo Bridge. It can be clearly seen that the effects of the unsteady aerodynamic drag force (P_i^*, $i = 2$, 3, 5 and 6) are significant. It was understood that this resulted from a large wind-induced torsional deflection of the deck up to 5–6° at its maximum. The effects became very significant particularly with the truss-stiffened deck of the Akashi-Kaikyo Bridge.

Important points to be noted on the flutter analysis are as follows:

1. A large wind-induced torsional deflection could occur for a long-span suspension bridge, particularly for a truss-stiffened long-span suspension bridge. The deflection, which is equivalent to giving the additional angle of attack, varies along the bridge axis, resulting in the change of deck cross-sectional shape against wind along the bridge axis. In addition, the aerodynamic interaction between cables and the deck exists. Therefore, the appropriate unsteady aerodynamic force coefficients (flutter derivatives) must be applied for the analysis considering the variation of the angle of attack for this case.
2. The unsteady aerodynamic lift and pitching moment have been usually considered in flutter analysis based on experience with the airfoil theory. However, it was found that the unsteady aerodynamic drag cannot be disregarded when the large wind-induced torsional deflection takes place, such as the case of the Akashi-Kaikyo Bridge. In particular, the unsteady aerodynamic drag resulting from the vertical and torsional motions play a role of destabilizing flutter. Therefore, the unsteady aerodynamic drag as well as the lift and pitching moment must be considered in such a case.

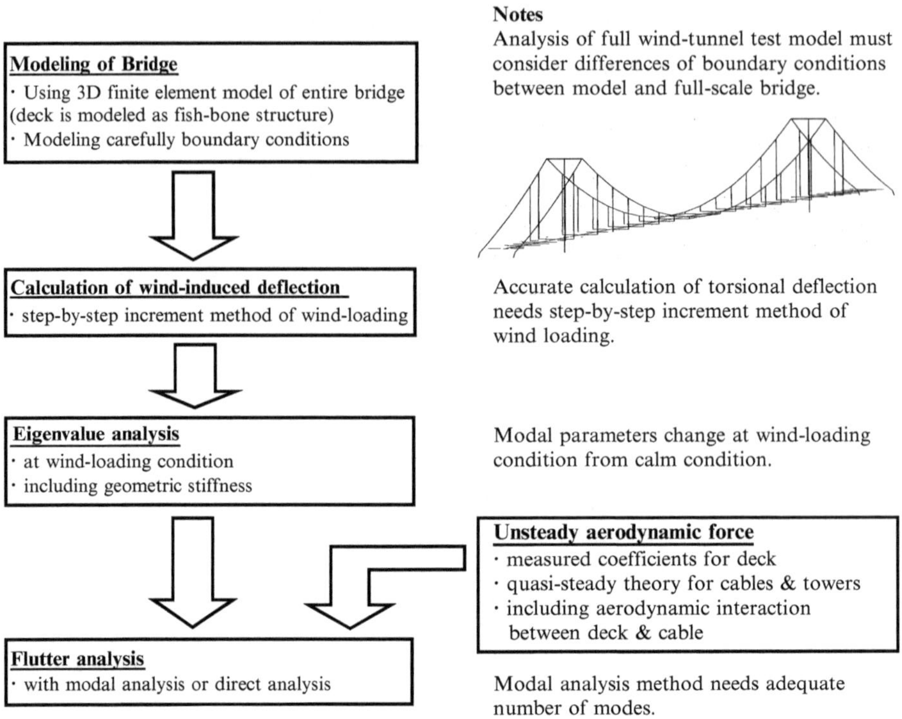

Fig. 7.2 Flow of flutter analysis recommended

3. Since the equilibrium with the wind loading introduces additional geometric stiffness, vibration modes may be different with and without the static wind load. Because the solution given by the modal analysis depends on the mode shapes chosen for the analysis, it is desirable that the mode shapes be obtained with the static wind load and they should be applied to flutter analysis.

   A flow chart of flutter analysis is shown in Fig. 7.2.

7.2.2 Other Applications of Flutter Analysis

There are some other applications of flutter analysis. They include a comparative study [25] of section model versus full bridge model tests of the Akashi-Kaikyo Bridge, an investigation [26] toward coupled flutter of a mono-cable suspension bridge, an investigation [27] regarding flutter of the Kurushima-Kaikyo Bridge and another hypothetical super long-span suspension bridge considered for the future long-span bridge projects [28].

As for the last application, a super long-span suspension bridge with the center span of 2,800 m and a 2-box deck was tested with its 1/125-scaled aeroelastic

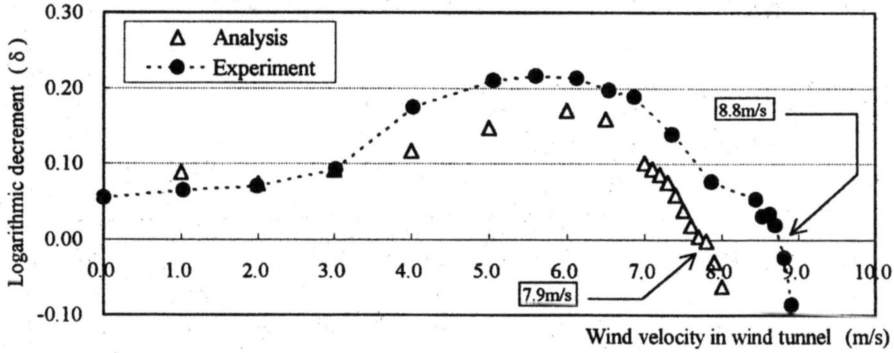

Fig. 7.3 Comparison between flutter analysis and full-model wind tunnel test of Super Long-span Bridge [28]

model. The test results were compared with the analytical results from the same method as that of the Akashi-Kaikyo Bridge. The flutter analysis agreed fairly well with the wind tunnel test results, as shown in Fig. 7.3.

7.3 Buffeting Analysis

As already described, buffeting analysis [2–4] of long-span bridges in Japan has been basically based on the random vibration theory in the frequency domain developed by Davenport [1]. This is due to the simplicity of analysis to obtain the necessary statistical values in design such as the maximum and mean. Buffeting analysis in the time domain is now utilized, but only the frequency domain analysis is specified in the design codes.

Based on the modal analysis method (7.4) described in the previous section, covariance ${\sigma_u}^2$ of displacement $\mathbf{u}$ and the power spectral density (PSD) $\mathbf{S}_{\Phi\Phi}$ of generalized coordinate $\mathbf{\Phi}$ are obtained as follows:

$$\sigma_u^2(i,j) = \int_0^\infty \mathbf{S}_{\mathbf{uu}} df = \int_0^\infty \mathbf{X}\,\mathbf{S}_{\Phi\Phi}\mathbf{X}^T df \tag{7.10}$$

$$\mathbf{S}_{\Phi\Phi} = \mathbf{E}^{-1}\mathbf{S}_{\mathbf{QQ}}[\mathbf{E}^{+}]^{-1} \tag{7.11}$$

where $\mathbf{E}(=\lambda^2\tilde{\mathbf{M}} + \lambda\tilde{\mathbf{C}} + \tilde{\mathbf{K}})$ is an impedance matrix, $\mathbf{S}_{\mathbf{QQ}}$ is PSD of the buffeting force, + as a superscript represents the conjugate transpose, and *i, j* indicate the points of interest.

Buffeting force consisting of lift L_b, drag D_b and pitching moment M_b, and resulting from longitudinal u and vertical w components of wind velocity fluctuations is expressed as

$$L_b = \frac{1}{2}\rho U^2 B\left\{C_L \chi_L^u \frac{2u}{U} + [C'_L + C_D]\chi_L^w \frac{w}{U}\right\} \tag{7.12a}$$

$$D_b = \frac{1}{2}\rho U^2 B\left\{C_D \chi_D^u \frac{2u}{U} + C'_D \chi_D^w \frac{w}{U}\right\} \tag{7.12b}$$

$$M_b = \frac{1}{2}\rho U^2 B^2\left\{C_M \chi_M^u \frac{2u}{U} + C'_M \chi_M^w \frac{w}{U}\right\} \tag{7.12c}$$

where C_L, C_D and C_M are the lift, drag and pitching moment coefficients, respectively. ($'$) represents their derivatives with respect to the angle of attack. χ is the aerodynamic admittance where the superscript and subscript indicate the input and output components, respectively.

By expressing the buffeting force as a linear superposition of both u- and w-components,

$$\mathbf{P}_b(t) = \{.....,L_b, D_b, M_b, ...\}^T = \mathbf{P}_b^u(t) + \mathbf{P}_b^w(t) \tag{7.13}$$

PSD of the buffeting force is expressed by

$$\mathbf{S_{QQ}} = \mathbf{X}^T\mathbf{S}_{P_iP_j}^{S_{uu}}\mathbf{X} + \mathbf{X}^T\mathbf{S}_{P_iP_j}^{S_{ww}}\mathbf{X} + \mathbf{X}^T[\mathbf{S}_{P_iP_j}^{C_{uw}} + i\mathbf{S}_{P_iP_j}^{Q_{uw}}]\mathbf{X} + \mathbf{X}^T[\mathbf{S}_{P_jP_i}^{C_{uw}} - i\mathbf{S}_{P_jP_i}^{Q_{uw}}]\mathbf{X} \tag{7.14}$$

where $\mathbf{S}_{P_iP_j}^{S_{uu}}$, $\mathbf{S}_{P_iP_j}^{S_{ww}}$, $\mathbf{S}_{P_iP_j}^{C_{uw}}$ and $\mathbf{S}_{P_iP_j}^{Q_{uw}}$ are the cross-spectral density functions of the fluctuating forces between points i and j, where the superscript represents an associated fluctuating wind velocity component. That is to say that S_{uu}, and S_{ww} are PSD of u- and w-components, respectively, and C_{uw} and Q_{uw} are the co- and quad-spectral density of the uw correlation, respectively.

Since there are aeroelastic coupling effects between the vibration modes, called the mode coupling, non-zero off-diagonal terms exist in the impedance (7.10) and (7.11) and buffeting force (7.14) matrices. The computer analysis, because of this non-symmetric nature, takes a long time and certain simplifications are usually taken. The methodology of buffeting analysis executed for the Honshu-Shikoku Bridges can be summarized as follows:

By neglecting the off-diagonal mode coupling terms, (7.11) can be decomposed into a SDOF equation of motion for each mode. In other words, PSD of generalized coordinate vector is expressed as the multiplication of PSD of the fluctuating wind speed, aerodynamic admittance, joint mode acceptance (spatial correlation) and the mechanical transfer function. The uw cross-spectrum in (7.14) is usually neglected to simplify the analysis and only PSDs of u- and w-components are incorporated. Hence

$$\mathbf{S}_{\Phi_m} = |\mathrm{E}|^2 \{\tilde{\mathbf{X}}_m\}^T \mathbf{R} \{\tilde{\mathbf{X}}_m\} |\chi|^2 \tilde{\mathbf{P}}_b^2 \mathbf{S}_v \tag{7.15}$$

where $\{\tilde{\mathbf{X}}_m\}$ is a column vector of deflections (vertical, lateral or torsion) reduced from the mth column vector of $\mathbf{X}$. $\mathbf{R}$ is the spatial correlation function, χ is the aerodynamic admittance function, $\tilde{\mathbf{P}}_b$ is a buffeting force coefficient defined below and S_v ($v = u$ or w) is PSD of u- or w-component.

The buffeting force of (7.12) can be simplified as

$$\tilde{\mathbf{P}}_b = \frac{1}{2}\rho UB[C'_L + C_D] \quad \text{(Lift)} \tag{7.16a}$$

$$= \rho UAC_D \quad \text{(Drag)} \tag{7.16b}$$

$$= \frac{1}{2}\rho UB^2 C'_M \quad \text{(Pitching moment)} \tag{7.16c}$$

Aerodynamic damping in the impedance matrix resulting from unsteady aerodynamic force effects, in which only those associated with vertical and lateral motion are considered, is defined based on the quasi-steady theory rather than flutter derivatives. Aerodynamic damping in torsion is set to zero to be conservative in design.

$$H_1^* = -\frac{1}{K}[C'_L + C_D] \tag{7.17}$$

$$P_1^* = -\frac{2C_D}{K}; \quad \text{others} = 0 \tag{7.18}$$

When the design of the Honshu-Shikoku Bridges was made, vibration modes were decomposed into three (vertical, lateral and torsional) components and analysis with (7.15) was executed for each component because of the efficiency of analysis. Therefore, the standard deviation of the fluctuating deflection to be evaluated was obtained by taking a square-root of the sum of the three-component responses of modes considered.

$$\sigma_{\mathbf{u}} = \sqrt{\sum_m^{\text{mode}} (\sigma_{\mathbf{u}}^L)^2 + \sum_m^{\text{mode}} (\sigma_{\mathbf{u}}^D)^2 + \sum_m^{\text{mode}} (\sigma_{\mathbf{u}}^M)^2} \tag{7.19}$$

where $\sigma_{\mathbf{u}}^L$, $\sigma_{\mathbf{u}}^D$ and $\sigma_{\mathbf{u}}^M$ are the standard deviation of fluctuating deflection in question, resulting from lift, drag and pitching moment, respectively. When the correlation among the components and/or the modes should be considered, formulation is given in a reference [29].

The expectation of the maximum deflection of **u** is finally obtained as a function of time T, assuming that **u** is a narrow-band stationary random process.

$$\mathbf{u}_{\max} = \mathbf{u}_{\text{mean}} + \sqrt{\sum_{m}^{\text{mode}} (g_m^L \cdot \sigma_{\text{u}}^L)^2 + \sum_{m}^{\text{mode}} (g_m^D \cdot \sigma_{\text{u}}^D)^2 + \sum_{m}^{\text{mode}} (g_m^M \cdot \sigma_{\text{u}}^M)^2} \tag{7.20}$$

where g_m^L, g_m^D and g_m^M are peak factors associated with the mode (m) and force components (L, D and M), defined as

$$g = \sqrt{2\ln \tilde{f}T} + \frac{0.5772}{\sqrt{2\ln \tilde{f}T}} \tag{7.21a}$$

$$\tilde{f} = \left[\int_0^{\infty} f^2 \mathbf{S_{uu}}(f)df \bigg/ \int_0^{\infty} \mathbf{S_{uu}}(f)df \right]^{1/2} \tag{7.21b}$$

Thanks to the progress in computer technology, the size of the analysis is not problematic any more. Analysis where a vibration mode is not decomposed into three but all mode couplings are incorporated, according to (7.10–7.14), can be now commonly utilized.

Expectation of the maximum internal reactions (bending moment or shear force) are finally obtained based on the maximum deflection **u**.

Defining the expectations of the mean and maximum of analytical section force as S_{mean} and $S_{\max}$, respectively, $S_{\max}/S_{\text{mean}}$ is called the gust response factor. In the wind resistant design codes [3, 4], this gust response factor is defined as a correction factor of the static wind load, and μ_2 and μ_3 are defined for a laterally large structure and a vertically large structure, respectively. Thus the design wind load P_D is calculated as

$$P_D = (\mu_2 \text{ or } \mu_3)\frac{\rho U^2}{2} C_D A \tag{7.22}$$

7.3.1 *Important Points in Buffeting Analysis*

Methodology of buffeting analysis was also verified by comparing its outcome with the results of full bridge model tests of the Akashi-Kaikyo Bridge. Important points found from this comparison are as follows:

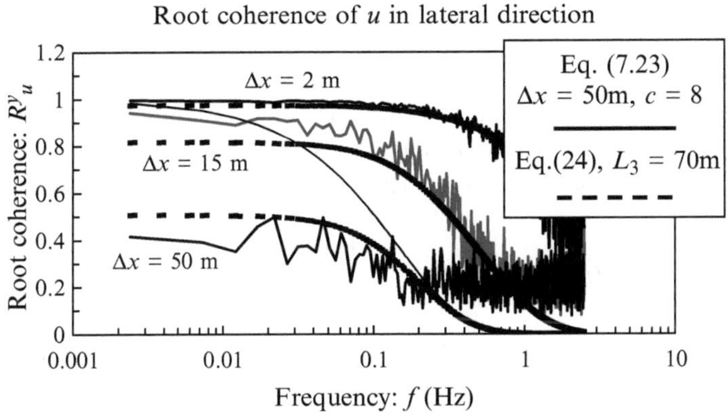

Fig. 7.4 Root coherence of longitudinal fluctuating wind velocity measured at separate locations in transverse direction in the wind tunnel test of the Akashi-Kaikyo Bridge

7.3.1.1 Spatial Correlation of Fluctuating Wind Speed

An exponential decay function has been assumed for a spatial correlation of the buffeting forces as shown in (7.23)

$$R(i,j,f) = \exp\left[-\frac{f}{U_i + U_j}\sqrt{(c_x\Delta x)^2 + (c_y\Delta y)^2}\right] \tag{7.23}$$

where c_x and c_y are the decay factors in lateral (Δx) and vertical (Δy) directions, respectively. Δx and Δy are spatial distance in lateral and vertical direction between points i and j.

However, it was pointed out in the full bridge model tests that the exponential decay function of (7.23) would overestimate the spatial correlation particularly for the low frequency components. In order to improve the accuracy of the function, an alternative function by Roberts & Surry (7.24), which was applied to the analysis of the Lions' Gate Bridge [30], was also used and gave a better result as shown in Fig. 7.4. For the convenience in analysis, not the co-coherence but root-coherence function is compared with (7.24).

For longitudinal component

$$R_u(\Delta x,f) = \frac{2^{1/6}}{\Gamma(5/6)}\left[\eta^{5/6}\mathrm{K}_{5/6}(\eta) - \frac{\eta^{11/6}}{2}\mathrm{K}_{1/6}(\eta)\right] \tag{7.24}$$

where $\eta = \frac{\Delta x}{L_3}B_1\sqrt{1 + \left(\frac{2\pi}{B_1}\right)^2\left(\frac{fL_3}{U}\right)^2}$, $B_1 = \sqrt{\pi}\frac{\Gamma(5/6)}{\Gamma(1/3)}$, Γ is the gamma function, $\mathrm{K}_{5/6}$ and $\mathrm{K}_{1/6}$ are the modified Bessel function of the second kind and L_3 is twice of the turbulence scale in transverse direction of longitudinal wind velocity fluctuations.

7.3.1.2 Aerodynamic Admittance

The definition of aerodynamic admittance function has been a focal point for many researchers. In the case of a line-like structure such as a long-span bridge, the Sears function developed for airfoils is sometimes used for aerodynamic admittance of lift due to vertical wind fluctuations (χ_L^w) approximated as shown in (7.25).

$$\left|\chi_L^w(f)\right|^2 = \frac{0.1811 + k}{0.1811 + (0.1811\pi + 1)k + 2\pi k^2} \tag{7.25}$$

On the other hand, the aerodynamic admittance of drag due to along-wind fluctuations (χ_D^u) is defined based on Davenport theory as shown in (7.26)

$$\left|\chi_D^u(f)\right|^2 = \frac{2}{(k_z\xi)^2}\left[k_z\xi - 1 + e^{-k_z\xi}\right]; \ \xi = \frac{fd}{U} \tag{7.26}$$

Unlike the case of airfoils, the aerodynamic admittance function for bridge decks cannot be theoretically obtained. It is, therefore, desirable to obtain it by wind tunnel tests, though admittedly it is an impossible task to measure the force correlation in an ideally "two-dimensional turbulence".

Aerodynamic admittance of the Akashi deck model was obtained experimentally [31]. The results showed that the results of lift and pitching components were different from the Sears function specified in the design codes. The reason of the difference is not clear. Estimation of aerodynamic admittance from the measured flutter derivatives was also attempted [32].

7.3.1.3 Mechanical Admittance (Mechanical Transfer Function)

The impedance matrix in (7.11) is influenced by the self-excited forces and has mode-coupling effects. In other words, all components of the matrix are non-zero terms. If the effect of the self-excited force can be ignored, the impedance becomes a diagonal matrix and can be decomposed into the mechanical admittance of a single-degree-of-freedom system for each mode.

Self-excited force effects have been incorporated into buffeting analysis based on the quasi-steady assumption. Studies [33, 34] have pointed out that the accuracy of buffeting prediction is largely influenced by the flutter derivatives employed in its calculation.

7.3.2 Applications of Buffeting Analysis

Examples of comparison between buffeting analysis, full bridge model tests and the field observation of the bridge are shown in this section.

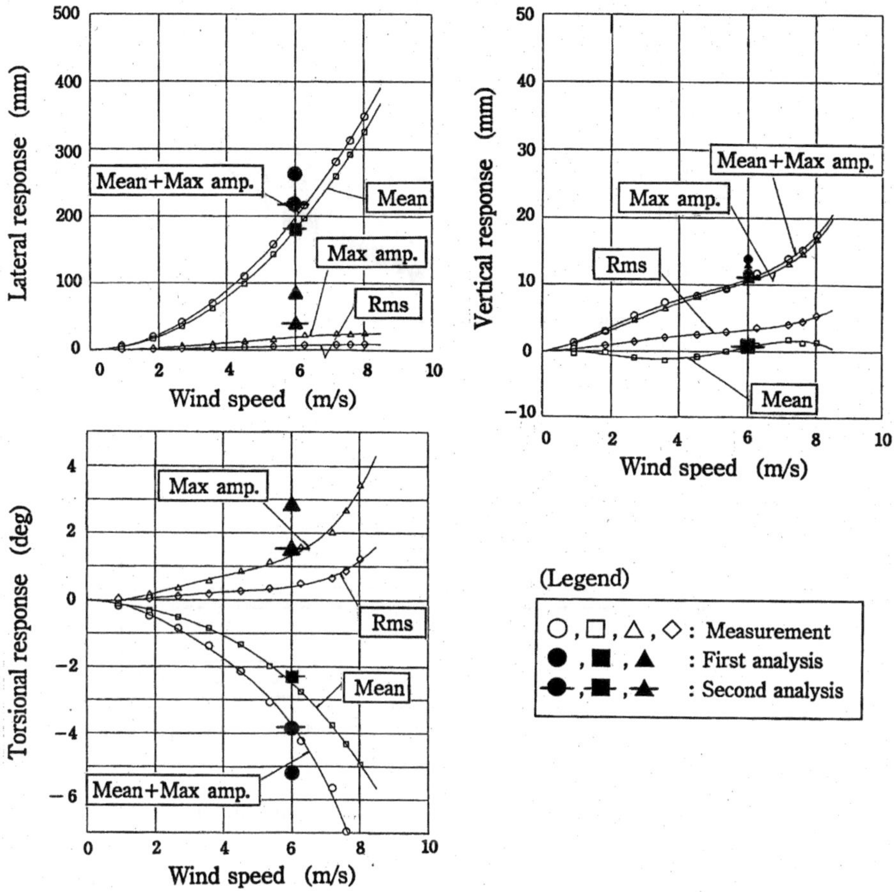

Fig. 7.5 Comparison of buffeting response between full bridge model tests and analysis

7.3.2.1 Full Bridge Model Tests of the Akashi-Kaikyo Bridge

Figure 7.5 shows a comparison [35] between the full bridge model wind tunnel tests and the buffeting analysis of the Akashi-Kaikyo Bridge. In the figure, the "First Analysis" represents the analysis based on the design code, where an exponential decay function is employed for the spatial correlation of the buffeting forces and the "Second Analysis" is the case where (7.24) was used as the spatial correlation. The "Second Analysis" also considered the torsional aerodynamic damping, which was disregarded in the design code. It is observed in the figure that the "Second Analysis" was able to demonstrate the full bridge model behavior very well.

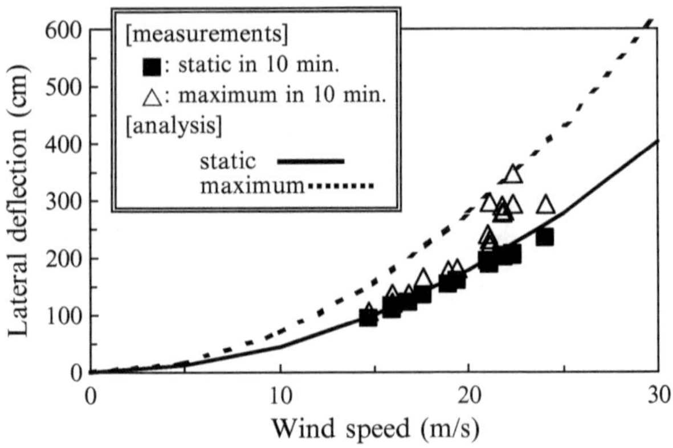

Fig. 7.6 Comparison of measured lateral deck deflection in typhoon wind with the design analysis. Wind was skewed less than 30° from normal to the bridge

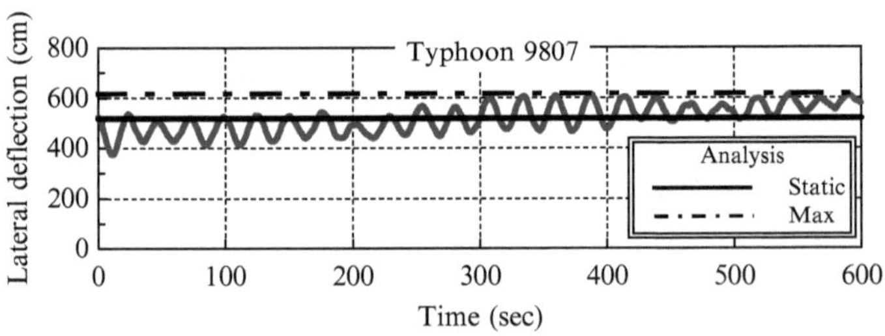

Fig. 7.7 Comparison of measured lateral deck deflection with buffeting analysis using the measured parameters

7.3.2.2 Field Monitoring of the Akashi-Kaikyo Bridge

Field monitoring [36, 37] of the Akashi-Kaikyo Bridge has been conducted since its completion. Lateral deflection of the bridge measured under the typhoon wind is compared with the analytical results in Fig. 7.6. A good agreement in terms of the static lateral deflection is observed between the measurement and prediction, whereas the measured dynamic deflection including the fluctuating component are found lower than predicted based on the design code [3].

Analysis [38] of data collected under some strong winds showed that the spatial correlation of wind speed fluctuation in natural wind was lower than what is given by the exponential decay function particularly in low frequency range (see Fig. 4.9).

Based on the parameters obtained during the monitoring, the measured data can be well simulated [38] by buffeting analysis as shown in Fig. 7.7. However, there is a large variation in the measured data themselves. Therefore, monitoring and analysis is being continued in order to make buffeting analysis more realistic.

7.3.2.3 Application of Buffeting Analysis to Other Bridges

The application of buffeting analysis and full-scale comparison was also attempted for other bridges of the Honshu-Shikoku link [39, 40], the Hakucho Bridge [41] and the Ikara-jima Bridge [42]. There are also similar studies [43–46] of a cable-stayed bridge.

In particular, for the case of the Hakucho Bridge, an extensive identification of higher vibration modes up to the 19th mode was successfully made by the ambient vibration survey. The data collection was made by densely located accelerometers and with the Random Decrement method and Ibrahim's Time Domain analysis [47, 48]. Furthermore, an inversion method was established in which the changes of stiffness and damping of the structure due to aerodynamic forces were inversely identified as equivalent stiffness and damping only from the obtained modal characteristics [48, 49].

7.4 Vortex-Induced Vibration

Vortex-induced vibration, in which the maximum response is of particular interest, has been widely investigated by wind tunnel tests. There are also some formulae [17, 18] for the prediction of dynamic response, based on wind tunnel test data, and they can be applied to a preliminary design as described in Sect. 2.2. More precise analytical prediction is seldom used because formulating of the aerodynamic forces in vortex-induced vibration is difficult. However, the Hanshin Expressway Public Corporation uses an analytical formula [18] of vortex-induced vibration as explained below.

Assuming that the vortex-induced vibration is a self-excited vibration, the aerodynamic force can be expressed as follows:

$$F_L = \frac{1}{2}\rho U^2 B C_h \dot{y}/U \tag{7.27}$$

where F_L is the lift force per unit deck length and C_h is the self-excited force coefficient.

By defining h as the structural critical damping ratio and assuming that the amplitude of the vortex-induced vibration is determined solely by the self-excited force of (7.27), the following relationship can be obtained:

$$h - \frac{1/2\rho U B C_h}{2\omega m} = 0 \tag{7.28}$$

Furthermore, assuming that the self-excited force coefficient C_h is inversely proportional to the non-dimensional amplitude y/B,

$$C_h = \frac{C}{y/B} \tag{7.29}$$

the amplitude of vortex-induced vibration y can be obtained from (7.28) and (7.29) as follows:

$$y = \frac{\rho U B^2 C}{4 m \delta f} \tag{7.30}$$

Once the self-excited force coefficient C can be obtained by wind tunnel tests, the amplitude of vortex-induced vibration can be thus analytically estimated. Actually C can be expressed by D/B (D is deck height and B deck width) based on the past experience as

$$C = 0.625(D/B)^2 \tag{7.31}$$

In addition, the wind speed where the amplitude of vortex-induced vibration reaches the maximum is given as $U = 1.67\alpha f B$ ($\alpha = 1.2$–1.8). Then, the amplitude of vortex-induced vibration is given by

$$y = \frac{K_R \alpha \rho D^2 B}{4 m \delta} \tag{7.32}$$

where K_R is a reduction factor that takes account of the effects of the wind turbulence.

References

1. Davenport AG (1962) The response of slender, line-like structures to a gusty wind. Proc Inst Civil Eng 20:389–408
2. Honshu-Shikoku Bridge Authority (1976) Wind resistant design standard for Honshu-Shikoku bridges (in Japanese). HSBA, Kobe
3. Honshu-Shikoku Bridge Authority (1990) Wind resistant design specification for Akashi-Kaikyo bridge (in Japanese). HSBA, Kobe
4. Honshu-Shikoku Bridge Authority (1994) Wind resistant design standard for Onomichi-Imabari route (in Japanese). HSBA, Kobe
5. Miyata T, Yamada H, Yokoyama K, Kanazaki T, Iijima T, Tatsumi M (1992) Construction of boundary layer wind tunnel for long-span bridges. J Wind Eng Ind Aerod 41–44:885–896
6. Miyata T, Yokoyama K, Yasuda M, Hikami Y (1992) Akashi Kaikyo bridge: wind effects and full model wind tunnel tests. In: Larsen A (ed) Aerodynamics of large bridges. Balkema, Rotterdam, pp 217–236
7. Miyata T, Yamaguchi K (1993) Aerodynamics of wind effects on the Akashi Kaikyo bridge. J Wind Eng Ind Aerod 48:287–315
8. Miyata T, Tada K, Sato H, Katsuchi H, Hikami Y (1994) New findings of coupled-flutter in full model wind tunnel tests on the Akashi Kaikyo Bridge. In: Proceedings of the cable-stayed and suspension bridges, Deauville 2:163–170
9. Miyata T, Kitagawa M, Yamada H, Kanazaki T, Toriumi R (1994) Design considerations on 3-dimensional gust response in the Akashi Kaikyo bridge. In: Proceedings of the cable-stayed and suspension bridges, Deauville 2:171–178

10. Miyata T (1995) Full model testing of large cable-supported bridges. In: Proceedings of the 9th ICWE, state-of-the-art volume, New Delhi, pp 249–280
11. Honshu-Shikoku Bridge Authority (1980) Wind-tunnel test specification for Honshu-Shikoku bridges (in Japanese). HSBA, Kobe
12. Scanlan RH (1978) The action of flexible bridges under wind, I: Flutter theory. J Sound Vib 60(2):187–199
13. Simiu E, Scanlan RH (1996) Wind effects on structures, 3rd edn. Wiley, New York
14. Jain A, Jones NP, Scanlan RH (1996) Coupled flutter and buffeting analysis of long-span bridges. J Struct Eng ASCE 122(7):716–725
15. Ito M (1998) 21st-Century super long span bridges in Japan. In: Larsen A, Esdahl S (eds) Bridge aerodynamics. Balkema, Rotterdam, pp 145–152
16. Ito M (1999) Large bridge projects in the far east. In: Miyata T, Fujisawa N, Yamada H (eds) Long-span bridges and aerodynamics. Springer, Tokyo, pp 40–55
17. Japan Road Association (1991) Wind resistant design manual for highway bridges (in Japanese). Maruzen, Tokyo. Also revised version in 2007
18. Hanshin Expressway Public Corporation (1984) Aerodynamic design method on wind resistant design (in Japanese). HEPC, Osaka
19. Miyata T, Yamada H (1988) Coupled flutter estimate of a suspension bridge. In: Proceedings of the international colloquium on bluff body aerodynamics and its application, Kyoto, pp 485–492
20. Miyata T, Yamada H, Kazama K (1995) On application of the direct flutter FEM analysis for long span bridges. In: Proceedings of the 9th ICWE, New Delhi, pp 1030–1041
21. Dung NN, Miyata T, Yamada H, Minh NN (1998) Flutter responses in long span bridges with wind induced displacement by the mode tracing method. J Wind Eng Ind Aerod 77&78:367–379
22. Wilde K, Fujino Y (1998) Aerodynamic control of bridge deck flutter by active surfaces. J Eng Mech ASCE 124(7):718–727
23. Wilde K, Fujino Y, Kawakami T (1999) Analytical and experimental study on passive aerodynamic control of flutter of bridge deck section. J Wind Eng Ind Aerod 80(1–2):105–119
24. Wilde K, Omenzetter P, Fujino Y (2001) Suppression of bridge flutter by active deck-flaps control system. J Eng Mech ASCE 127(1):80–89
25. Fujino Y, Iwamoto M, Ito M, Hikami Y, Takeda K, Miyata T, Tatsumi M (1991) Wind tunnel experiment of long-span suspension bridge using 3D models (Wind-induced response under smooth flow and its prediction) (in Japanese). J Wind Eng JAWE 48:1–17
26. Tanaka H, Yamamura N, Shiraishi N (1993) Multi-mode flutter analysis and 2- and 3-dimensional model tests on bridges with non-analogous modal shapes. J Struct Mech Earthquake Eng JSCE 471(I–24):35–46
27. Nakao T, Kusuhara S, Yamamura N, Tanaka H, Kobayashi Y, Hatanaka H (1994) Coupled flutter behavior of suspension bridges with flat box stiffening girder (in Japanese). In: Proceedings of the 13th national symposium on wind engineering, Tokyo, pp 389–394
28. Sato H, Hirahara N, Fumoto K, Hirano S, Kusuhara S (2001) Full aeroelastic model test of a super long-span bridge with slotted box girder. In: Proceedings of the 5th APCWE, Kyoto, pp 469–472
29. Kiureghian AD (1980) Structural response to stationary excitation. J Eng Mech Div ASCE 106(EM6):1195–1213
30. Irwin HPAH (1977) Wind tunnel and analytical investigations of the response of Lions' Gate bridge to a turbulent wind. Laboratory Technical Report, NRC, Canada
31. Sato H, Matsuno Y, Kitagawa M (1994) Evaluation of the aerodynamic admittance for the stiffening girder of the Akashi Kaikyo bridge (in Japanese). In: Proceedings of the 13th national symposium on wind engineering, Tokyo, pp 131–136
32. Hatanaka H, Tanaka H (2008) Aerodynamic admittance functions of rectangular cylinders. J Wind Eng Ind Aerod 96(6–7):945–953

33. Katsuchi H, Jones NP, Scanlan HR, Akiyama H (1998) A study of mode coupling in flutter and buffeting of the Akashi-Kaikyo bridge. Struct Eng/Earthquake Eng JSCE 15(2):175s–190s
34. Katsuchi H, Jones NP, Scanlan HR (1999) Multi-mode coupled flutter and buffeting analysis of the Akashi-Kaikyo bridge. J Struct Eng ASCE 125(1):60–70
35. Kitagawa M, Kanazaki T, Katsuchi H (1995) Study on gust response of Akashi-Kaikyo Bridge through full model wind tunnel test (in Japanese). Honshi Technical Report BOEA 19(75): 17–22
36. Honshu-Shikoku Bridge Authority (1998) The Akashi-Kaikyo bridge: design and construction of the world's longest bridge. HSBA, Kobe
37. Kashima S, Okano S, Takeguchi M, Mori K (2000) Monitoring system for the Akashi Kaikyo Bridge. In: Proceedings of the workshop on research and monitoring of long span bridges, Hong Kong, pp 119–126
38. Toriumi R, Katsuchi H, Furuya N (2000) A study on spatial correlation of natural wind. J Wind Eng Ind Aerod 87:203–216
39. Tatsumi M, Hata K (1990) Behavior of suspension bridges under the strong wind (in Japanese). Honshi Technical Report BOEA 14(53):2–11
40. Katsuchi H, Tada K, Kitagawa M (1998) A study on wind-resistant design of long-span bridges based on field observation result. Struct Eng/Earthquake Eng JSCE 15(1):115s–125s
41. Nakamura S (2000) GPS measurement of wind-induced suspension bridge girder displacements. J Struct Eng ASCE 126(12):1413–1419
42. Niihara Y, Hayashida K, Nakano R, Kubo Y (1999) Full scale measurement of wind pressure on a deck of a concrete cable-stayed bridge, Part3: Turbulent scale effects on the gust force (in Japanese). J Wind Eng JAWE 79:245–257
43. Kimura K, Nakamura S, Tanaka H (1994) Buffeting analysis for cable-stayed bridges during construction in Yawed wind. In: Proceedings of the cable-stayed and suspension bridges, Deauville 2:109–116
44. Katsuchi H, Kitagawa M, Sato H, Kato M (1995) Study on analytical method for gust response of long-span cable-stayed bridge (in Japanese). J Struct Eng JSCE 41A:793–800
45. Hatanaka A, Ogasawara M, Katsuchi H, Yamamura N (1995) Coherence for buffeting response analysis of long-span bridges (in Japanese). J Construct Steel JSSC 3:115–122
46. Matsuda K, Toriumi R, Iwasaki T, Tokushige M (1998) Horizontal gust response of a long-span cable-stayed bridge under construction (in Japanese). Proceedings of the 15th national symposium on wind engineering, Tokyo, pp 377–382
47. Niiyama M, Sato M, Ikeda K, Sugawara T, Sato K (2001) Applicability of micro-tremors measuring method for estimation of natural vibration frequencies of Hakucho-Ohashi (in Japanese). J Struct Eng JSCE 47A:469–477
48. Abe M, Fujino Y, Nagayama T, Ikeda K (2001) Identification of non-proportionally damped system using ambient vibration measurement and its application to a suspension bridge (in Japanese). J Struct Mech Earthquake Eng JSCE 689(I-57):261–274
49. Nagayama T, Abe M, Fujino Y, Ikeda K (2005) Structural identification of a non-proportionally damped system and its application to a full-scale suspension bridge. J Struct Eng ASCE 131(10):1536–1545

Chapter 8
Vibration Control for Bridge Girders

8.1 Wind-Induced Vibration and Its Control of the Bridge Girders

8.1.1 Wind-Induced Vibration of Bridge Girders

Wind-induced vibration of bridge girders needs to be considered as a part of the wind resistant design procedure in order to avoid unfavorable wind actions such as the collapse of the original Tacoma Narrows Bridge in 1940. The wind-induced vibration of bridge girders can be classified into the following three types [1]:

(a) Self-excited vibrations (galloping and flutter)
(b) Vortex-induced vibration
(c) Buffeting

For the box-girder sections with a small width-to-depth ratio, B/D (B: width, D: depth), for example, possibility of both galloping and flutter has to be examined. Since the self-excited vibrations are often catastrophic to the structure, it is usually more appropriate to entirely eliminate any possibility of having them by modifying the cross-sectional shape of the deck rather than simply reducing the magnitude of vibration.

Vortex-induced vibrations are further divided into two types. One is caused by the Kármán vortices and the other by the vortices generated along the deck section and is strongly affected by the movement of the bridge deck. Most of the vortex-induced vibrations observed with the flat box girders of cable-supported bridges are the latter type [2]. The magnitude of vortex-induced vibration is often influenced by wind turbulence. Figure 8.1 shows how the dynamic response amplitude depends on the flow turbulence [3]. In general, the larger the turbulence intensity is, the smaller the amplitude is for most of the box girder sections. However, the peak response amplitude for the case of the hexagonal box girders is found even greater

Y. Fujino et al., *Wind Resistant Design of Bridges in Japan*,
DOI 10.1007/978-4-431-54046-5_8, © Springer 2012

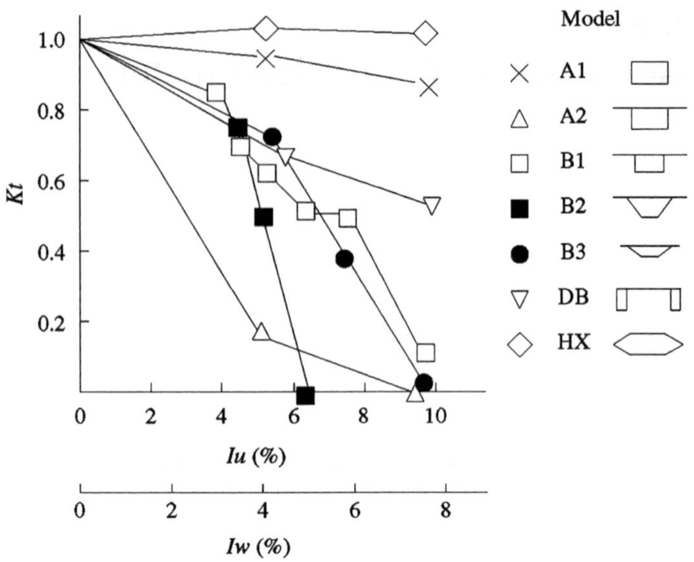

Fig. 8.1 Relationship between the turbulent intensity and the maximum amplitude ratio [3]

in turbulent flow than in smooth flow. These effects of turbulence are reflected in the *Wind Resistant Design Manual for Highway Bridges* in Japan as a reduction factor due to turbulence. Aerodynamic countermeasures against vortex-induced vibrations such as the addition of fairings and flaps have been widely employed. TMD (Tuned Mass Damper) is a typical mechanical countermeasure, which has been adopted since 1985 in Japan. For example, the TMDs were installed inside of the box girder of the Kansai National Airport Access Bridge (1991) [4].

Buffeting is a random vibration caused by the fluctuating wind load due to natural wind turbulence and its predominant action is in direction of the mean wind.

8.1.2 Design Procedure Against Wind-Induced Vibrations

8.1.2.1 Self-Excited Vibrations

Self-excited vibrations such as flutter and galloping can be catastrophic to the bridge and hence the perfect control of them have to be considered at the design of the bridge. First, the reference wind speed is to be defined by considering the design wind speed and wind turbulence. Next, it needs to be ascertained that the critical wind speed of the bridge, under the worst condition, is still higher than the reference speed. A typical wind resistant design of bridges in Japan has been done with the help of section model wind tunnel tests in smooth flow, since it allows the use of a relatively large model. Detailed discussion regarding the choice of testing methods is found in Chap. 6.

8.1.2.2 Vortex-Induced Vibration

Vortex-induced vibration of a bridge girder does not cause a catastrophic damage immediately. Hence, the control of it can be decided after the unacceptable vibration is really observed during the bridge construction or even after its completion. The reduction factor due to turbulence, which was explained in relation to Fig. 8.1, is often considered for many aerodynamically bluff sections, with the exception of the flat hexagons. It is required that either the critical wind speed for vortex-induced vibration is higher than the design wind speed or the amplitude of vortex-induced vibration is small enough to be allowed. The allowable amplitude is decided based on the consideration of the first passage failure, fatigue damage, or serviceability. Generally the allowable amplitude is decided based on the serviceability of the bridge. The amplitude equivalent to the acceleration of 1.00 m/s^2 is adopted in the *Wind Resistant Design Manual for Highway Bridges.*

8.1.2.3 Buffeting

Buffeting of a bridge girder is taken into design consideration, not directly as the dynamic response but as the increment of static wind load. For this reason, buffeting is not usually examined as its allowable amplitude. In exceptional cases, however, where the first passage, fatigue damage, or serviceability has to be taken into account, the buffeting amplitude is checked against the allowable amplitudes obtained in a similar manner as for vortex-induced vibration.

8.1.3 Various Methods to Control Wind-Induced Vibrations

8.1.3.1 Aerodynamic Countermeasures

Practice of aerodynamic countermeasures in Japan [1, 5–8] to control the wind-induced vibrations of bridge girders is briefly explained in this section. Their function is to control the flow pattern around the bridge girder by changing the cross-sectional shape so that the aerodynamic forces would be reduced.

As it is mentioned in the previous section, three different kinds of wind-induced vibrations can be considered for bridge girders. Although it is difficult to come up with any specific countermeasures for each of the three aerodynamic phenomena, the following general discussion may be possible. The vibration control methods that have been practiced in Japan are as listed in Tables 8.1–8.5.

Table 8.1 Aerodynamic countermeasure for suspension bridges

H: Heaving, T: Torsional
1) mass per unit span length [t/m] or [tm^2/m]

No.	Bridge Name	Form span[m]	Structural Condition	Mass 1)	Freq. [Hz]	Deck Section Shape	B/D
1	Akashi-Kaikyo (1998) [28.Hyogo]	3-span 2-hinge truss-stiffened suspension bridge 960+1991+960	H T	43.79 5,336	0.064 0.150	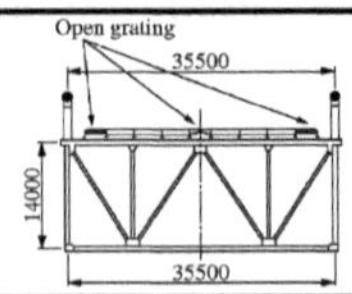	2.54
2	Akinada (1999) [34.Hiroshima]	3-span 2-hinge suspension bridge 255+750+170	H T		0.164 0.503	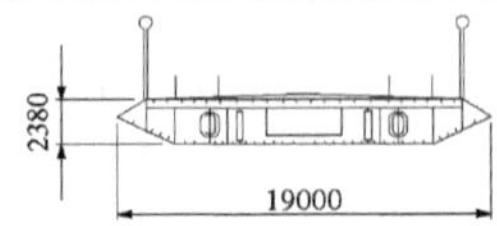	7.98
3	Hakuchou (1998) [1.Hokkaido]	3-span 2-hinge rigid box girder suspension bridge 330+720+330	H T	14.8	0.12 0.45	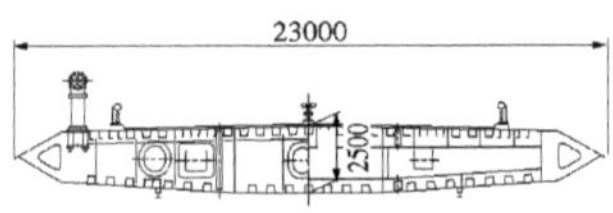	9.2
4	Innoshima (1983) [34.Hiroshima]	3-span 2-hinge truss-stiffened suspension 250+770+250	H T	21.05 2,231	0.178 0.374	D=9m 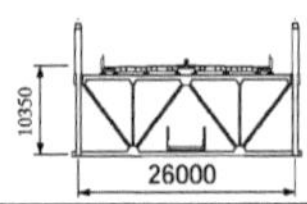	2.89
5	Kurushima (1999) [38.Ehime]	2-span 2-hinge suspension bridge 250+1020 1030	H T		0.15 0.37	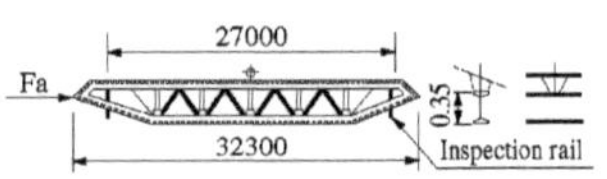	7.5
6	Ohnaruto (1984) [36.Tokushima]	3-span 2-hinge suspension bridge 330+876+330	H T		0.155 0.308	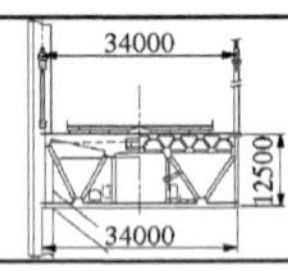	2.72
7	Rainbow (1993) [13.Tokyo]	3-span 2-hinge truss-stiffened suspension 114+570+114	H T		0.168 0.413	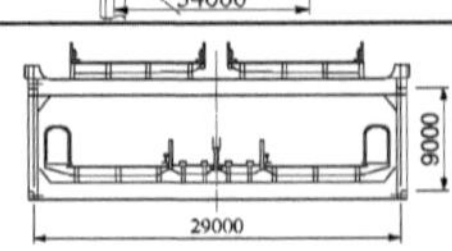	3.26
8	Seishi hananohashi (1984) [47.Okinawa]	1-span suspension bridge	H T	1.0 2.386	0.93 0.64	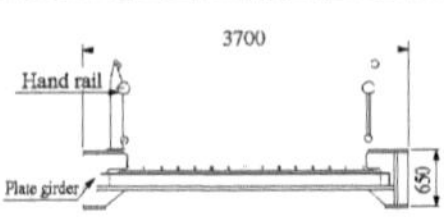	5.7
9	Shiki (1979) [32.Shimane]	1-span box-stiffened suspension bridge 150	H T	1.978 1.490	0.647 3.749	B=3.8m 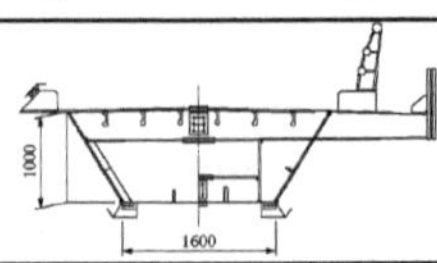	3.80
10	Wakato (1962) [40.Fukuokaq]	3-span 2-hinge truss-stiffened suspension 89+367+89	H T			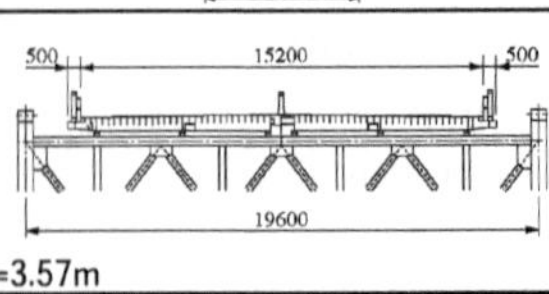 D=3.57m	5.49

Fa: Fairing, Fl: Flap, D: Deflector,
HP: Horizontal Plate, VP: Vertical Plate,
Sk: Skirt, Sp: Spoiler

α: Attack angle
H: Heaving, T: Torsional
V: Vortex induced vibration
F: Flutter, G: Galloping
Gu: Gust response

A: First Passage
B: Fatigue
C: Serviceability
D: Erection instrument safty
N: no information

Device								Phenomenon	Effect						Requisite Condition	
Shape	Fa	Fl	D	HP	VP	Sk	Sp		Flow	α[deg.]	w/o Device Amplitude	w/o Device Vel. [m/s]	with Device Amplitude	with Device Vel. [m/s]	Value	Reason
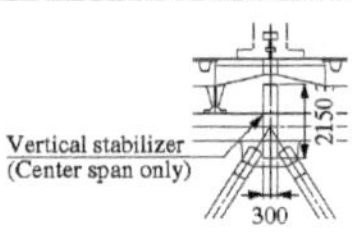					○			T-F	S (west)	0		60		83	78m/s	Vref
									S (east)	0		60		89	78m/s	Vref
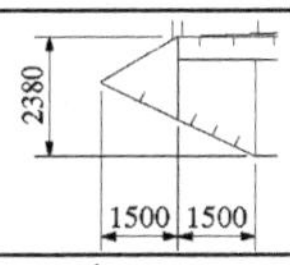	○							no data available								
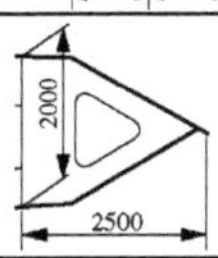	○							T-F	S	-3		-		88	> 67	N
										0		-		91	> 67	
										3		-		88	> 67	
										5		-		79	> 37	
										(3D-model)					[m/s]	
Stopping block 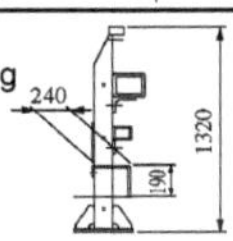			○					T-F	S	0				>100	66m/s	Vref
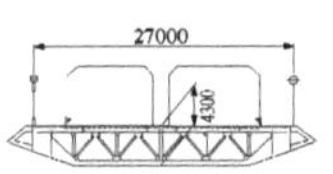	○							T-F	S	0				88 ↑	70	A
								T-F	S	3				72 ↑	64	A
										(2D-model)						
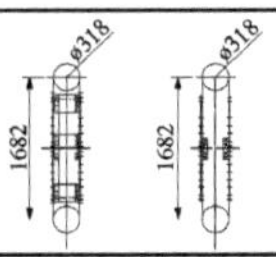					○			no data available								
					○											
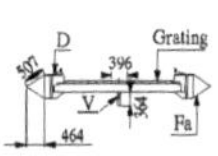	○		○		○			F	S	0		80		7	77	N
										-5		80 ↑		—	[m/s]	
										5		77 ↑		8		
								H−V	S	0	0	7	37	7		N
										-5	55	5	26	6		
										5	55	8				
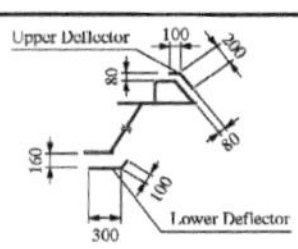			○					H-G	S	0		34		>70	66m/s	Vref
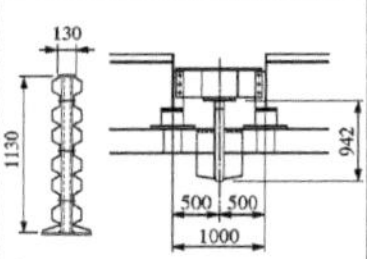					○			T-F	S	0		40		>80		N

Table 8.2 Aerodynamic countermeasure for cable-stayed bridges

H: Heaving, T: Torsional
1) mass per unit span length [t/m] or [tm²/m]

No.	Bridge Name	Form span[m]	Structural Condition	Mass 1)	Freq. [Hz]	Deck Section Shape	B/D
1	Denden (1980) [1.Hokkaido]	3-span continuous cable-stayed bridge 60+120+30	H T	3.80 N	0.98 2.55	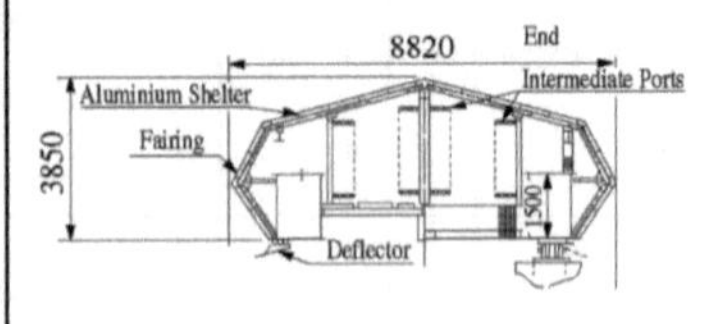	2.3
2	Eisai (1977) [14.Kanagawa]	2-span continuous steel cable-stayed bridge 144+90	H T	7.566 89.03	0.555 2.669	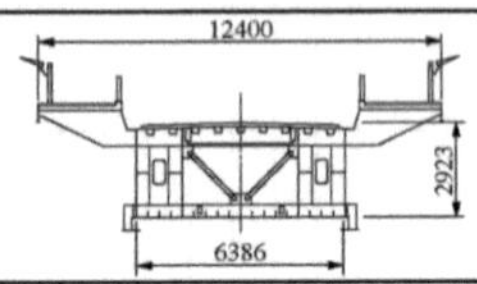	4.24
3	Fujito・Hinoura (1980) [28.Hyogo]	2-span continuous steel cable-stayed bridge 120+30	H T	4.9 44.2	0.75 1.70	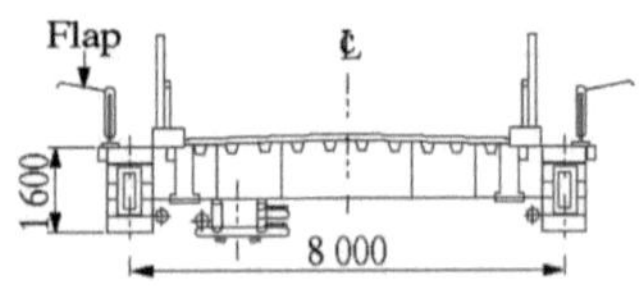	5.0
4	IshikariKako (1975) [1.Hokkaido]	3-span continuous steel cable-stayed bridge 64+160+64	H T	- 		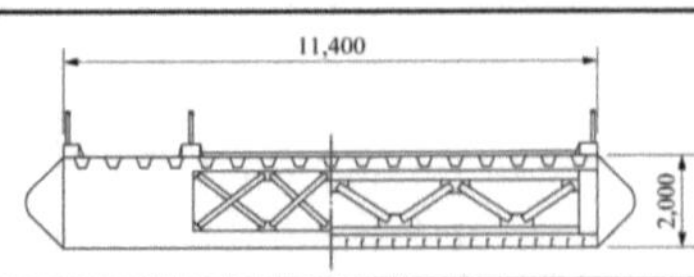	
5	Kamome (1975) [27.Osaka]	3-span continuous steel cable-stayed bridge 100+240+100	H T	N	N	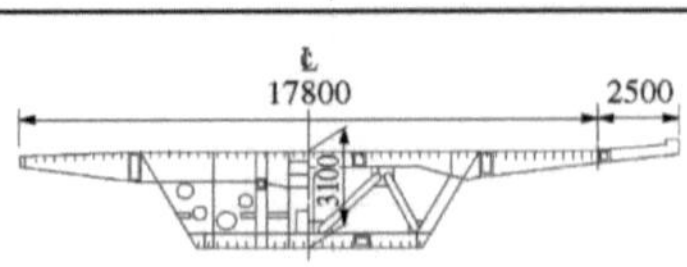	5.7
6	Katsushika Harp (1986) [13.Tokyo]	4-span continuous steel cable-stayed bridge 40+134+220 +60	H T	13.1 -	1.15 1.52	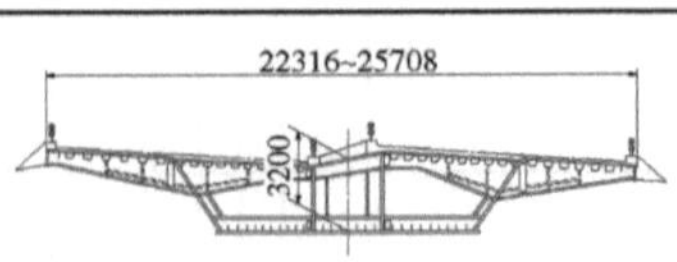	7.0~ 8.0
7	Kosoku Komatsugawa (1970) [13.Tokyo]	3-span continuous steel cable-stayed bridge 60+160+60	H T	N	N	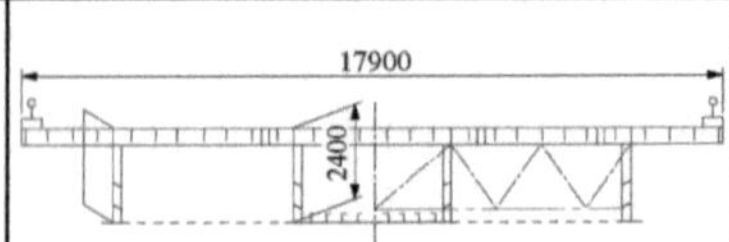	7.25

Fa: Fairing, Fl: Flap, D: Deflector,
HP: Horizontal Plate, VP: Vertical Plate,
Sk: Skirt, Sp: Spoiler

α: Attack angle
H: Heaving, T: Torsional
V: Vortex induced vibration
F: Flutter, G: Galloping
Gu: Gust response

A: First Passage
B: Fatigue
C: Serviceability
D: Erection instrument safty
N: no information

Device								Phenomenon	Effect						Requisite Condition	
Shape	Fa	Fl	D	HP	VP	Sk	Sp		Flow	α[deg.]	w/o Device		with Device		Value	Reason
											Amplitude	Vel. [m/s]	Amplitude	Vel. [m/s]		
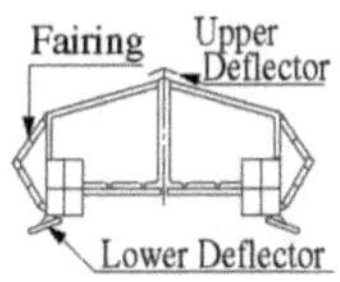			○					H-V	S	0	250 [mm]	18	39 [mm]	24		
								T-V	S	0	2 [deg.]	45	1.2 [deg.]	63		
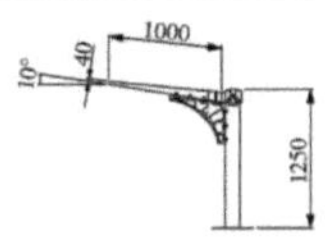		○						H-V	S	0	124 [mm]	15-18	8 [mm]	16.4		N
								H-G	S	0		38		>64	50m/s	Vref
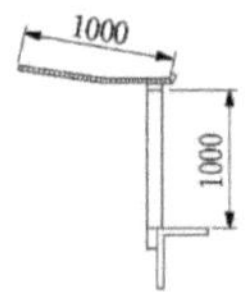		○						H-V	S	0	no-comparision		84 [mm]	11		
													48 [mm]		(d=0.03)	
								T-V	S	0			0.8 [deg.]	31.9		N
	○							H-V	-		[mm]		[mm]			
											no data available					
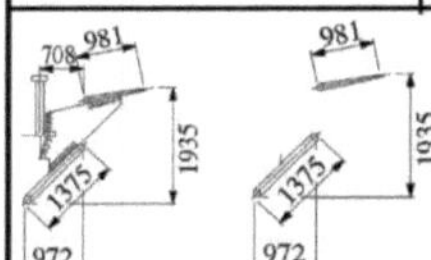		○									no data available					
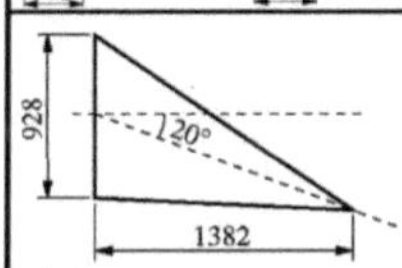	○							H-V	-	5	610 [mm]	28	30 [mm]	28		N
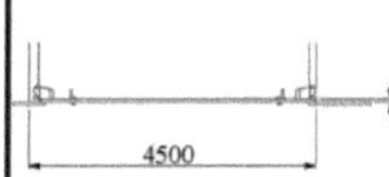				○ (overplate)				H-V	S	+5	500 [mm]	22	340 [mm]	20.2		N
										+2	208 [mm]	20.8	0	—		
								T-F	S	+5	diverge	75.6	diverge	80.0		N

(continued)

Table 8.2 (continued)

H: Heaving, T: Torsional
1) mass per unit span length [t/m] or [tm^2/m]

No.	Bridge Name	Form span[m]	Structural Condition		Freq. [Hz]	Deck Section Shape	B/D
				Mass 1)			
8	Megami [42.Nagasaki]	3-span continuous steel cable-stayed bridge 200+480+200	H T		0.25 0.67	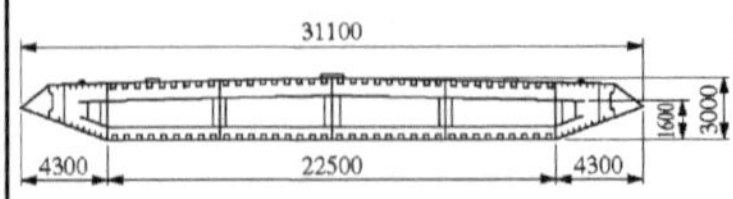	10.4
9	Meiko Higashi (1997) [23.Aichi]	3-span continuous steel cable-stayed bridge 145+410+145	H T		0.336 1.865	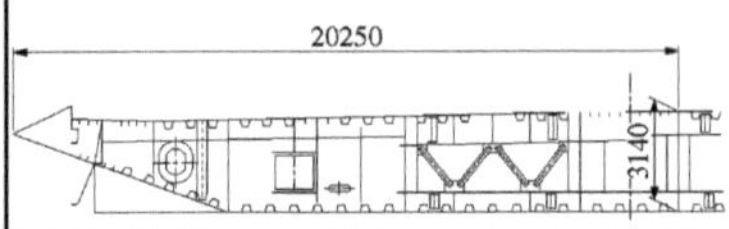	11.94
10	Meiko Nishi (1985) [23.Aichi]	3-span continuous steel cable-stayed bridge 175+405+175	H T			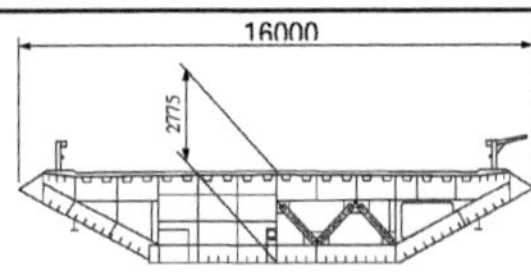	5.77
11	Okutama (Umezawa) (1988) [13.Tokyo]	2-span continuous steel cable-stayed bridge 105+160	H T	13.2 309.1	0.462 1.932	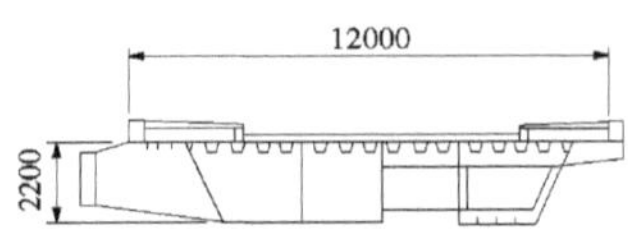	5.57
12	Orizuru (1980) [27.Osaka]	2-span continuous steel cable-stayed bridge 67+34	H T	10.19 555.7 (model)	1.316 7.878	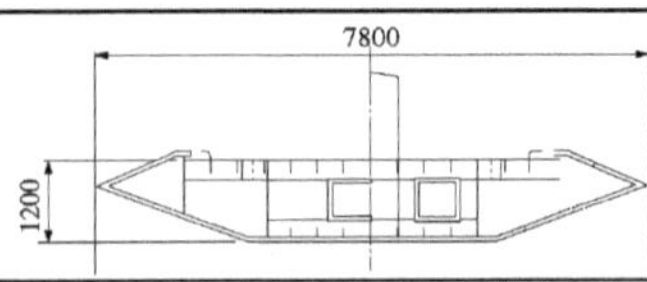	6.5
13	Ohshiba (1997) [34.Oshiba]	3-span continuous PC cable-stayed bridge 100+210+100	H T		0.340 1.047	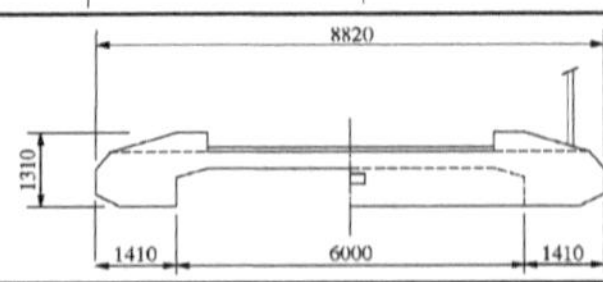	6.73
14	Sugawara Shirokita (1988) [27.Osaka]	3-span continuous steel cable-stayed bridge 119+238+119	H T	N	N	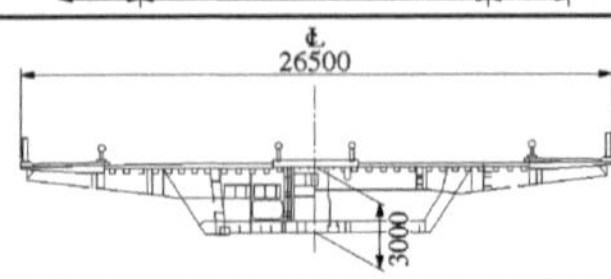	8.83

Fa: Fairing, Fl: Flap, D: Deflector,
HP: Horizontal Plate, VP: Vertical Plate,
Sk: Skirt, Sp: Spoiler

α: Attack angle
H: Heaving, T: Torsional
V: Vortex induced vibration
F: Flutter, G: Galloping
Gu: Gust response

A: First Passage
B: Fatigue
C: Serviceability
D: Erection instrument safty
N: no information

Device								Phenomenon	Effect						Requisite Condition	
Shape	Fa	Fl	D	HP	VP	Sk	Sp		Flow	α[deg.]	w/o Device		with Device		Value	Reason
											Amplitude	Vel. [m/s]	Amplitude	Vel. [m/s]		
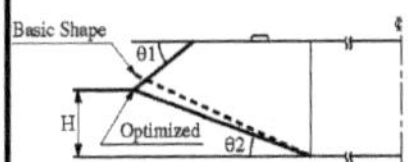 θ1=38.6deg. θ2=20.4deg. H=1600mm	○				○*			H-V	S	3	187	8	-	-		
*: Spacing of inspection cart rails was optimized.								H-V	S	5	140	8	-	-		
											[mm]		[mm]			
								T-V	S	3	0.6	17	-	-		
								T-V	S	5	0.6	17	-	-		
								T-V	S	5	0.3	32	0.3	17		
											[deg.]		[deg.]			
										(2D-model)						
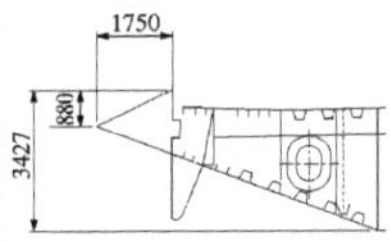	○										no data available					
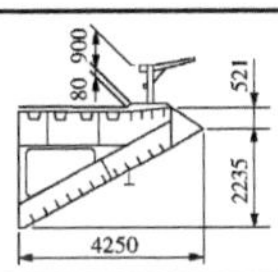	○	○						H-V	S	0			6	6		N
													[mm]			
								T-F	S	0				>79	66m/s	Vref
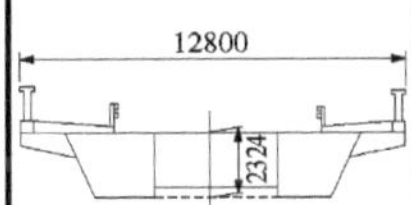				○				H-V	S	-3	85	N	100	N		
overplate										0	82.5	N	91	N		
										+3	224	N	1735	N		
										+5	310	N	278	N		
											[mm]		[mm]			
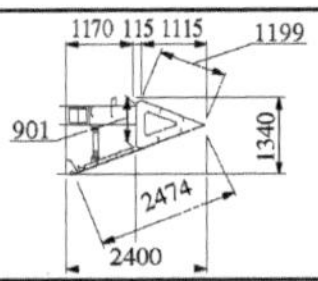	○							H-V	S	+4	67	15	7.5	9.6		N
											[mm]		[mm]			
								T-V	S	+4	1.4	38	0.45	49		N
											[deg.]		[deg.]			
								T-F	S	+4	diverg	45	0	—		N
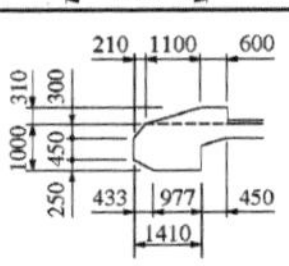	○							T-F	S	0, ±2				>80	67m/s	Vref
										+4				>80	43.75	Vref
										+6				>80	31.25	Vref
										-4				>62	43.75	Vref
										-6				>56	31.25	Vref
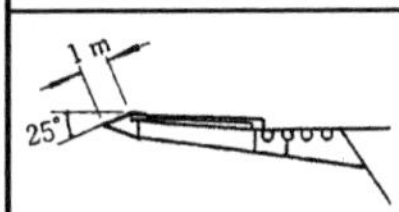	○							H-V	S	5	400	15	-	-		
											[mm]					
								T-V	S	5	240	25	-	-		
											[mm]					
								T-F	S	5	diver-gence	65		>80		

(continued)

Table 8.2 (continued)

H: Heaving, T: Torsional
1) mass per unit span length [t/m] or [tm^2/m]

No.	Bridge Name	Form span[m]	Structural Condition			Deck Section Shape	B/D
				Mass 1)	Freq. [Hz]		
15	Takanashi (1983) [32.Shimane]	2-span continuous complex cable-stayed bridge 99.5+57.5	H T	3.4 ~ 5.7 8.5 ~ 11.0	0.72 2.50	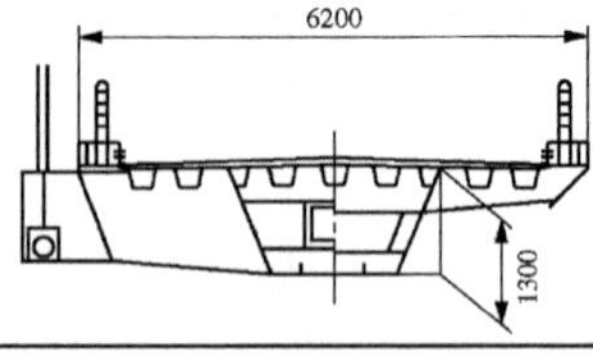	3.7
16	Tatara (1999) [38.Ehime]	3-span continuous composite cable-stayed bridge 270+890+320	H T			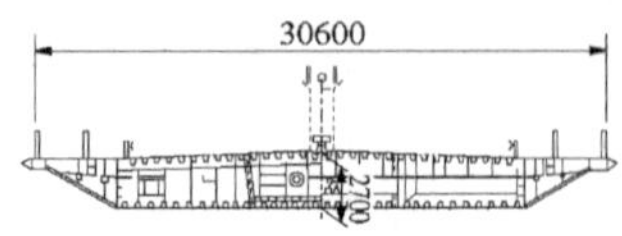	11.3
17	Tenpozan (1988) [27.Osaka]	3-span continuous steel cable-stayed bridge 179+350+120	H T	24.63 2,278	0.330 1.163	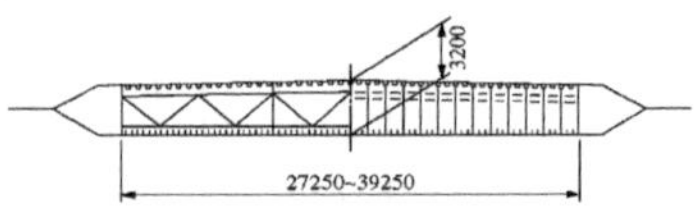	8.5- 12.27
18	Toda Koen (1987) [11.Saitama]	2-span continuous steel cable-stayed bridge 45+134	H T	2.889 144.9	1.04 6.05	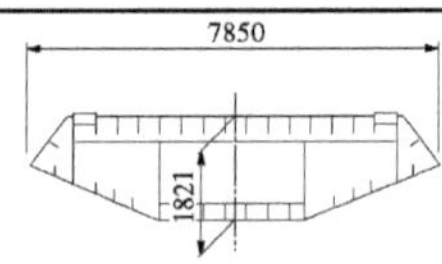	4.3
19	Tokachi Chuo (1988) [1.Hokkaido]	3-span continuous steel cable-stayed bridge 100+250+100	H T			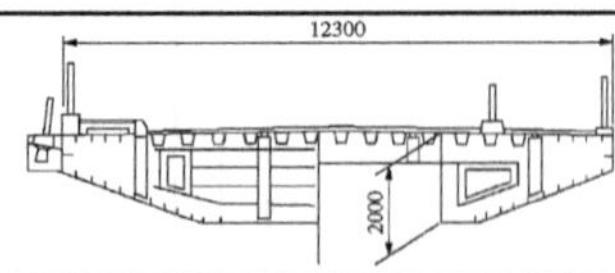	6.15
20	Torigai Niwaji (1987) [27.Osaka]	3-span continuous steel cable-stayed bridge 124.8+200 +61.3	H T	18.37 1310	0.85 1.9	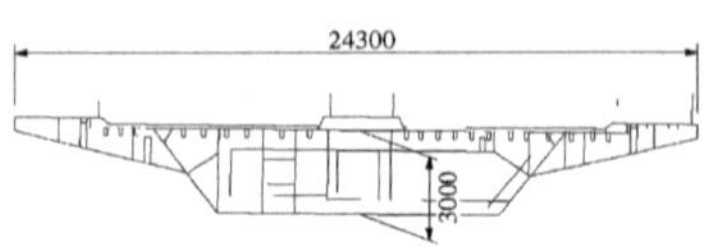	8.1
21	Tsurumi Tsubasa (1994) [14.Kanagawa]	3-span continuous steel cable-stayed bridge 255+510+255	H T	~ ~	0.21 0.55	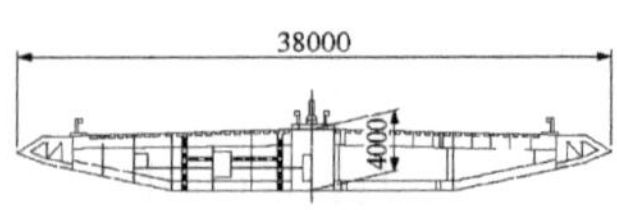	9.5
22	Yokohama Bay Bridge (1989) [14.Kanagawa]	3-span continuous steel cable-stayed bridge 200+460+200	H T	~		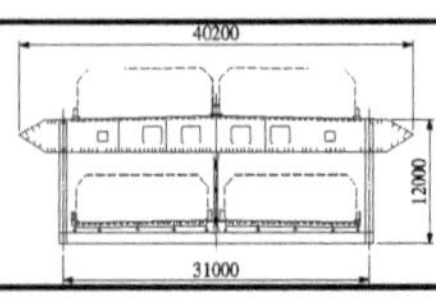	3.4

Fa: Fairing, Fl: Flap, D: Deflector,
HP: Horizontal Plate, VP: Vertical Plate,
Sk: Skirt, Sp: Spoiler

α: Attack angle
H: Heaving, T: Torsional
V: Vortex induced vibration
F: Flutter, G: Galloping
Gu: Gust response

A: First Passage
B: Fatigue
C: Serviceability
D: Erection instrument safty
N: no information

Device Shape	Fa	Fl	D	HP	VP	Sk	Sp	Phenomenon	Flow	α[deg.]	Effect w/o Device Amplitude	w/o Device Vel. [m/s]	with Device Amplitude	with Device Vel. [m/s]	Requisite Condition Value	Requisite Condition Reason
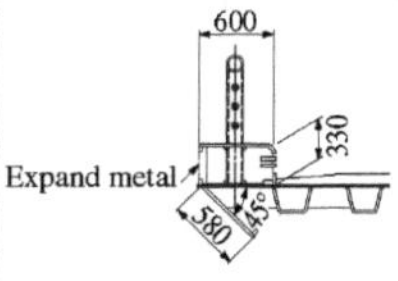	Slant plate at girder edge & Wheel guard with mesh panel							H-V	S	0	220 [mm]	15	1 [mm]	15		N
								T-V	S	0	2.1 [deg.]	27	0.0 [deg.]	27		
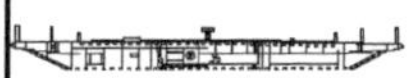	○										no data available					
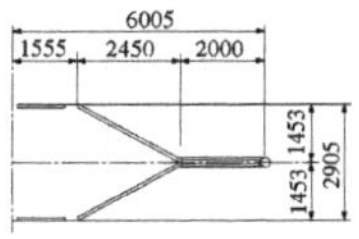				○				H-V	S	+7			-			
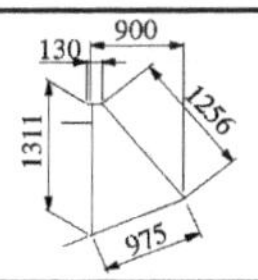	○							H-V	S	-3	no-comparision		134 [mm] (1DOF Mode)	12		N
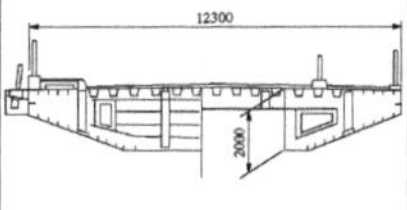	○							H-V	S				-		100 [mm]	C
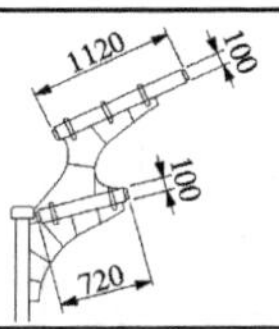		○						H-V	S	+5	1000 [mm]	29	0	-		N
								T-V	S	+5	0.86 [deg.]	41	0	-		N
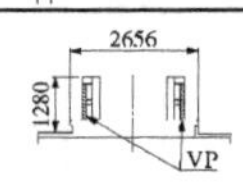 VP is for parallel bridge in the future	○				○			F	S	0		110↑		110↑		
										3		98.4↑		105↑		
								F 【in parallel】	S	0		108		110↑		
										3		83		96↑		
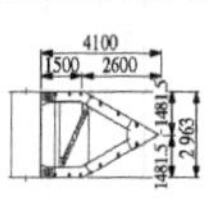	○							F	S	0				100↑	70	A
								H-V	S	0			0 [mm]			
										3			350 [mm]			

Table 8.3 Aerodynamic countermeasure for other type bridges

H: Heaving, T: Torsional
1) mass per unit span length [t/m] or [tm^2/m]

No.	Bridge Name	Form span[m]	Structural Condition	Mass 1)	Freq. [Hz]	Deck Section Shape	B/D
1	Ariake Nishi (1993) [13.Tokyo]	4-span continuous box girder bridge 92.5+125.5 +230+136	H T	N N	0.46 1.00	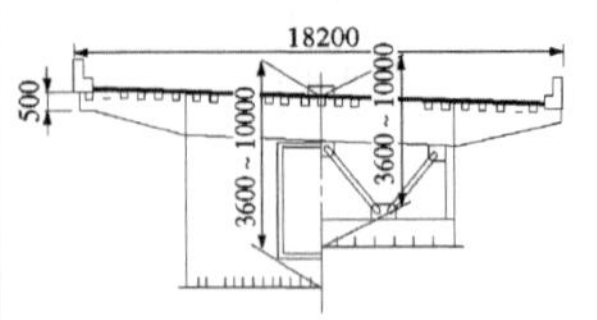	1.8~ 5.1
2	Kaita (1990) [34.Hiroshima]	3-span continuous steel box-girder bridge 148.5+250+148.5	H T	17.54 424.94	0.343 2.778	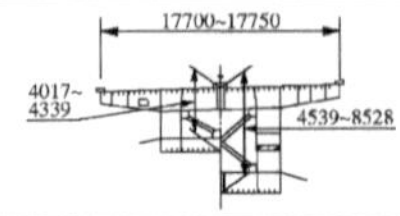	2.08- 4.42
3	Kansai (1991) (Roadway) [27.Osaka]	3-span continuous box girder bridge	H T	10.00 N	0.85 N	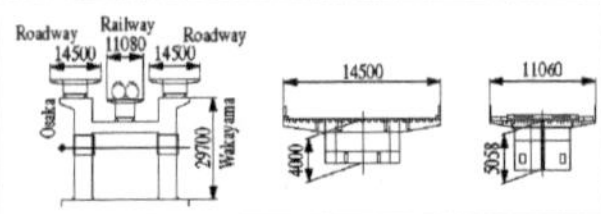	3.6
4	Muya (1985) [36.Tokushima]	4-span continuous steel plate-girder bridge 107+160+160+ 107	H T	8.790 N	0.492 N	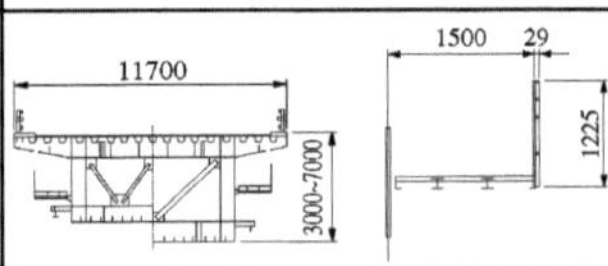	1.7~ 3.9
5	Namihaya (1995) [27.Osaka]	3-span continuous steel box-girder bridge 169.25+250 +159.35	H T			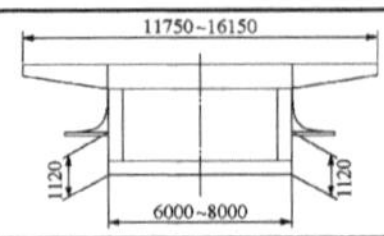	3.36
6	Shinkitakyushu Kuko [40.Fukuoka]	3-span continuous steel bridge stiffened with an arch at the 95+210+95	H T	15.60 (asymmetrical first mode) (symmetrical first mode 556	0.669 1.067 2.695	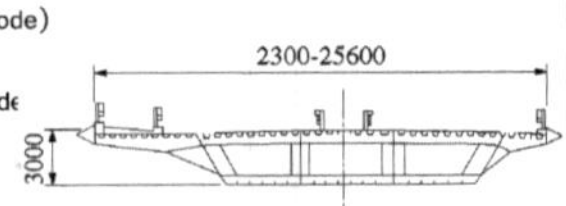	7.67 ~ 8.53
7	Shinkizugawa (1992) [27.Osaka]	4-span continuous steel plate-girder bridge 98+107.5 +107.5+107	H T			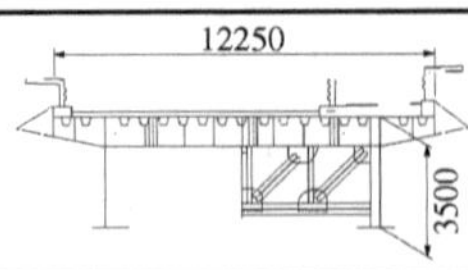	3.5
8	Tomari (1985) [47.Okinawa]	3-span continuous steel bridge 91.2+170+91.2	H T	16.1 ~ 17.0 580 ~ 600	0.56 2.15	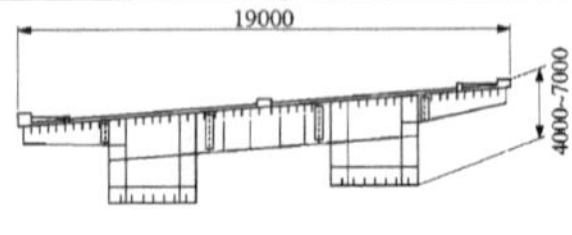	2.1~ 3.0
9	Tozaki (1983) [28.Hyogo]	4-span continuous steel cable-stayed bridge 149.6+190.4 +190.4+149.6	H	12.9	0.40	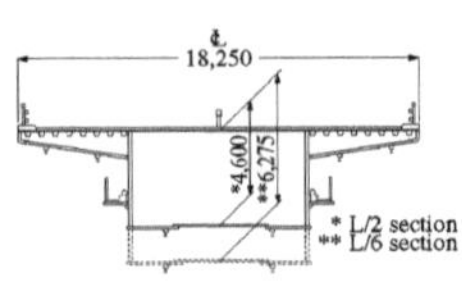	2.9~ 3.7

Fa: Fairing, Fl: Flap, D: Deflector,
HP: Horizontal Plate, VP: Vertical Plate,
Sk: Skirt, Sp: Spoiler

α: Attack angle
H: Heaving, T: Torsional
V: Vortex induced vibration
F: Flutter, G: Galloping
Gu: Gust response

A: First Passage
B: Fatigue
C: Serviceability
D: Erection instrument safty
N: no information

Device: Shape	Fa	Fl	D	HP	VP	Sk	Sp	Phenomenon	Flow	α[deg.]	w/o Device: Amplitude	w/o Device: Vel. [m/s]	with Device: Amplitude	with Device: Vel. [m/s]	Requisite Condition: Value	Requisite Condition: Reason
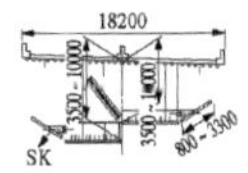						○		G	S	3		40		80↑		
								H-V	S	3	700 [mm]	20	500 [mm]	18		
										(2D-model)						
								G	S	0		35		80↑		
								H-V	S	0	600 [mm]	15		580 [mm]		
									T(7%)					400 [mm]		
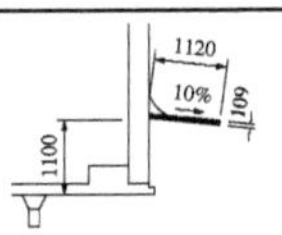				○				H-V	S	0	0.0113 [η/B]	15.4	0.0149 [η/B]	14.2		
								H-G	S	0		43.2		>100	65.5 [m/s]	Vref
											Section Model Test					
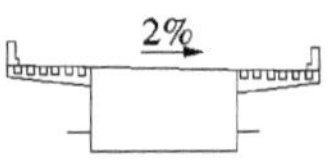				○				H-V	S	0			440 [mm]	20	(d=0.03)	
									S	0			300 [mm]	20	(d=0.06)	
									T(10%)	0			300 [mm]	20	(d=0.03)	
										(3D-model)						
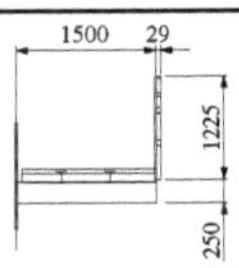					○			H-V	S	+5	386 [mm]	15				N
										0	211 [mm]	15	no-comparision			N
										-5	82 [mm]	18				N
								H-G	S	+5	liverge	28				N
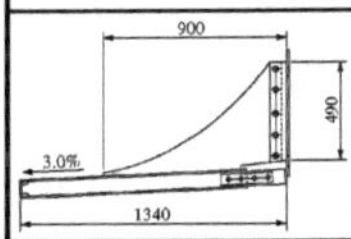				○				H-V	S	0		20	0		108mm	N
								H-G	S	0		20		>80	76.8	Vref
								V.I.V. occurs, followed by galloping continuously, as the wind speed increases.					'Device' includes Flap and skirt as well in the wind			
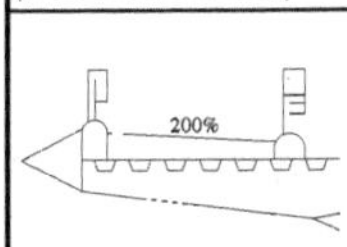	○							H-V	S	+4.5	56	38			(left wind)	
										+4.5	100	40			(right wind)	
										-4.5	0 [mm]	—	no-comparision		(right wind)	
								T-V	S	+4.5	1.3	85			(left wind)	
										+4.5	1.1	80			(right wind)	
										-4.5	0 [deg.]	-			(right wind)	
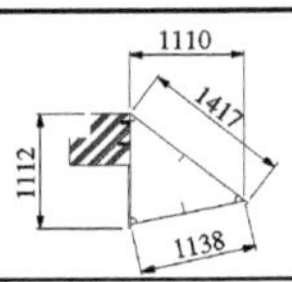	○	○		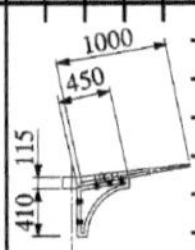									no data available			
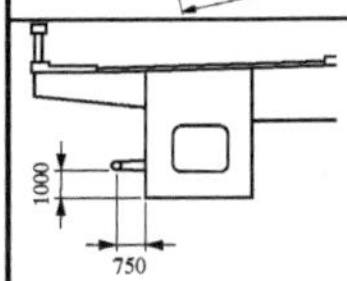				○				H-G	S	0		63		>97		N
										(L/6 Section)						
									S	0		62		>80		
										(3D-model)						
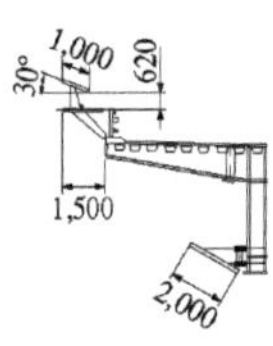		○				○		H-V	S	10	680 [mm]	17	40 [mm]	17	200 [mm]	N
										(L/2 Section)						
									S	0	1100 [mm]	17	250 [mm]	17		
										(3D-model with the terrain model)						
								H-G	S	10		30		-	> 92 [m/s]	N
										(L/2 Section)						
									S	0		35		>110		
										(3D-model with the terrain model)						

Table 8.4 Mechanical countermeasure for cable-stayed bridges

H: Heaving, T: Torsional
1) mass per unit span length [t/m] or [tm²/m]

No.	Bridge Name	Form	Structural Condition			Deck Section	
		span [m]		Mass 1)	Freq. [Hz]	Shape	B/D
1	Chichibu bridge (1985) [11.Saitama]	1-span steel box cable-stayed bridge 153	H T			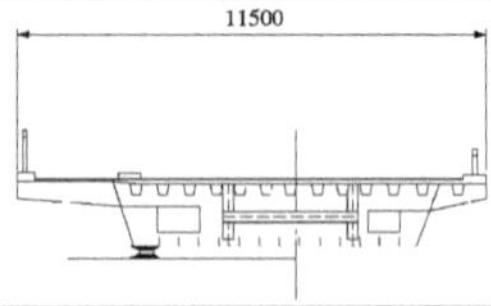	7.52
2	Okutama (Umezawa) (1988) [13.Tokyo]	2-span continuous steel cable-stayed bridge 105+160	H T	13.2 309.1	0.462 1.932	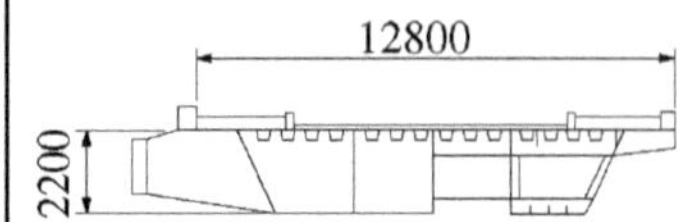	5.57

Flutter

Torsional flutter and coupled flutter can occur in bridge girders with relatively large width-to-depth ratio (*B*/*D*). One of the mechanisms to induce these instabilities is a pressure difference between the upper and lower side surface of the girder. Considering this fact, replacing a part of the deck floor by open grating is an effective way that decreases the pressure difference across the deck floor and could change the adverse flow pattern around the cross-section. In case of the shallow box girders whose width-to-depth ratio is small, various attachments such as fairings (Fa), wind noses, deflectors (D) and spoilers (Sp) can improve the flutter stability, because they can control the flow at the leading edge of the girder. In case of a truss girder, gratings on the edge of the deck can stabilize the flutter instabilities, because of the changing flow pattern around the deck and the upper chord. Moreover, the vertical stabilizer and the center barrier on a truss girder are known to be effective to reduce its instabilities, because the devices can weaken the action of the separation babbles from the leading edge.

There is another aerodynamic method using mechanical devices to decrease aerodynamic forces without changing the original girder shape. For example, there is a method of stabilizing the deck by controlling the motion of wings attached at the edge of the girder. By this means, the aerodynamic forces on the small flat plates installed on the girder edge may cancel the aerodynamic forces on the girder, and thus the flutter can be avoided. Moreover, moving a flap or fairing at the deck edge with some phase difference with the deck motion is found to reduce the wind-induced vibration such as flutter by modifying the aerodynamic characteristics [9, 10]. In addition, there are methods which can stabilize the flutter by installing a moving

Md: mass of the damper
μ: mass ratio
δ d: structural damping (logarithmic decrement)

α: Attack angle
H: Heaving, T: Torsional
V: Vortex induced vibration
F: Flutter, G: Galloping
Gu: Gust response 2) Total damping

A: First Passage
B: Fatigue
C: Serviceability
D: Erection instrument safty
N: no information

Device	Phenomenon	Effect								Requisite Condition	
		Flow	α[deg.]	w/o Device			with Device			Value	Reason
				δ	Amplitude	Vel. [m/s]	δ 2)	Amplitude	Vel. [m/s]		
TMD Md=0.4[t] δd=0.19	H-V	no data available TMD for girder was the firstly applied to Chichibu Bridge in Japan. (but, this is used only during the erection.									
TMD Md=6.4-9.6[t] (m=0.005-0.075) δd=0.4-0.5	H-V	S			-	-	0.12 (design)	-	-	8 [gal] d>=0.12	C

vertical plate on the deck edge. The phase of flow separation from the leading edge is controlled for this case, with the consideration of the phase characteristics of the wind-induced vibration. These active control methods using mechanical devices are still at the research stage and have not been applied in reality so far.

Galloping

Galloping is a divergent vibration, which occurs in bridge girders with relatively small width-to-depth ratio (*B*/*D*). The flow field under the bridge deck seems to be an important factor to control this phenomenon; the galloping is often avoided by changing the configuration of the girder cross-section, particularly on its lower side. As an aerodynamic countermeasure, the skirt at the lower corner of the girder with some inclined angle from vertical can reduce the wind velocity of the separated flow from the leading edge on the lower side surface, and thus the self-excitation force tends to be eliminated. In this case, the optimum length and inclined angle of the plate need to be chosen empirically by performing wind tunnel tests. In addition, the addition of narrow horizontal plates (Hp) equipped on the main girder web can also control the separated flow from the leading edge and can contribute to the stabilization.

As for mechanical approaches to stabilize galloping and even vortex-induced vibrations, there are methods to control the boundary layer separation from the frontal edge by using a rotor at the corner of rectangular cylinder or supplying an artificial air flow to the inner side of the boundary layer. Moreover, there has been

Table 8.5 Mechanical countermeasure for other type bridges

H: Heaving, T: Torsional
1) mass per unit span length [t/m] or [tm^2/m]

No.	Bridge Name	Form	Structural Condition			Deck Section	
		span[m]		Mass 1)	Freq. [Hz]	Shape	B/D
1	Parallel viaducts of Fukuoka Kitakyushu Urban Expressway [40.Fukuoka]	parallel 3-span continuous 91+113+89	H H H H (general mass)	821.6t 785.8t 832.9t 897.9t	0.707 0.713 1.026 1.071	1st mode(β=180deg) 1st mode(β=0deg) 2nd mode(β=180deg) 2nd mode(β=0deg) β: phase between two viaduc 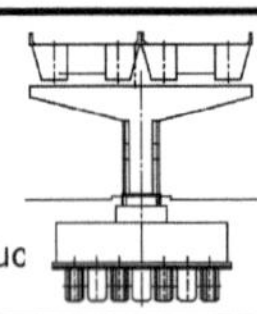	
2	Kansai International Airport Access (1991) [27.Osaka]	continuous box girder bridge (90+98)+(109+109)+(109+109+109)	H T		0.774 N	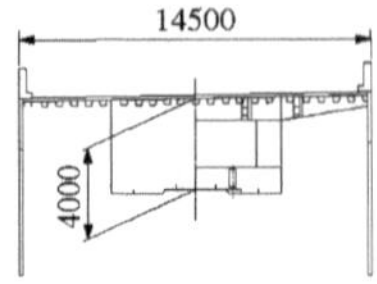	3.625
3	Namihaya (1995) [27.Osaka]	3-span continuous steel box-girder bridge 169.25+250+159.35	H T			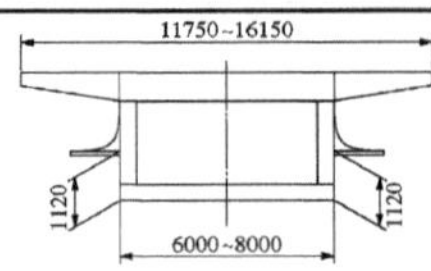	3.36
4	Oshima Bridge approaching viaduct (1999) [42.Nagasaki]	2-span continuous steel box-girder bridge 72.5+72.5 (Terashima Side) 95+65 (Otawa S.)	H H		1.224 (Terashima) 1.008 (Otawa)	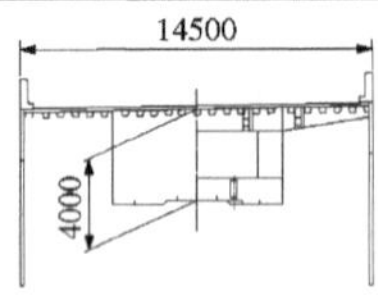	3.63
5	Tokyo Bay Aqua-Line (1997) [14.Kanagawa]	10-span continuous steel box 130+140+190+240+240+190+140+130+120+110=1630	H T	25.3 ~ 25.4 1249 ~ 1356	(1st) 0.329 (2nd) 0.471 1.52	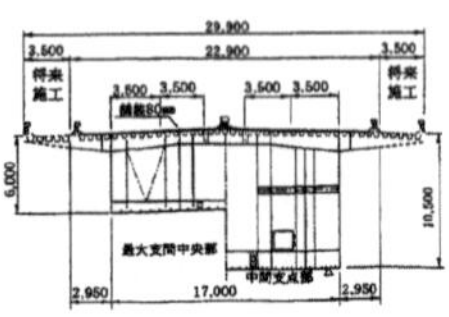	3.82
6	Chubu International Airport (U.C.) [23.Aichi]	2*5-span continuous steel box (Parallel 87+4@100 & 90+4@100 (P2-P7 portion	H	9.4 ~ 9.7	(1st) 0.69 (2nd) 0.73 (3rd) 0.83	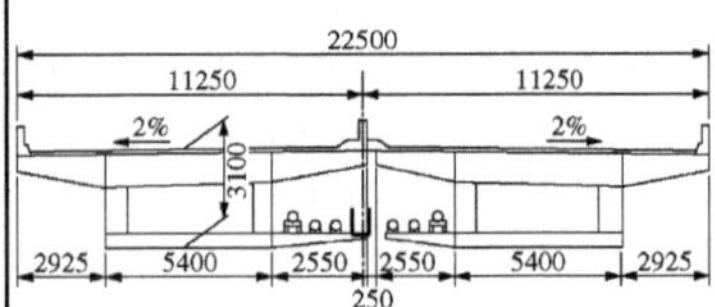	3.63

Md: mass of the damper
μ: mass ratio
δ d: structural damping (logarithmic decrement)

α: Attack angle
H: Heaving, T: Torsional
V: Vortex induced vibration
F: Flutter, G: Galloping
Gu: Gust response 2) Total damping

A: First Passage
B: Fatigue
C: Serviceability
D: Erection instrument safty
N: no information

Device	Phenomenon	Flow	α[deg.]	Effect						Requisite Condition	
				w/o Device			with Device			Value	Reason
				δ	Amplitude	Vel. [m/s]	δ 2)	Amplitude	Vel. [m/s]		
TMD	H-V	S T	0 0								
TMD	H-V	S	0	0.03	0.037B	28	0.06 with 26 TMDs, δ>0.1 was measured on site	0.017B	28		
TMD Three TMDs were installed. Two were for 1st mode at the center span (vertical & horizontal directions). The third one was for 3rd mode at the side span for vertical mode.	H-V										
Oil Damper											
Terashima Side mode 1	H-V	S	0	-	>50cm	27	0.062			δ >0.061	
mode 2	H-V	S	0	-			0.087			δ >0.068	
Otawa Side mode 1	H-V	S	0	-	>50cm	22	0.047			δ >0.049	
mode2	H-V	S	0	-			- (measured on the site)			δ >0.055	
TMD for 1st mode Md=80[t] (μ=0.012) δd=0.11	H-V (1st)			0.028 - 0.044 (measured on the site)	191 [gal]	17.8	0.22 (design)	25 [gal] (measured on the site)	17.5	50 [gal] δ>=0.22	C
TMD for 2nd mode Md=80[t] (μ=0.010) δd=0.11	H-V (2nd)			0.031 - 0.047 (measured on the site)	-	-	0.14 (design)	-	-	δ>=0.1	B
TMD	H-V (1st)	S	0		- (Galloping)	-	0.21 (design)	350	28		A,B
	H-V (2nd)	S	0		-	-	0.21 (design)	390	30		A,B
	H-V (3rd)	S	0		-	-	0.08 (design)	290	43		A,B

an attempt to control the flow separation at the leading edge and to reduce the vibration by applying artificial sound to the air flow around the bridge deck [11].

Vortex-Induced Vibration

Vortex-induced vibration also tends to take place when the bridge girders have a relatively small width-to-depth ratio (*b*/*d*) just like the case of galloping. The aerodynamic countermeasure, for this case, should be installed to control the flow separation from the leading edge. For example, the fairings (Fa) tend to promote the reattachment of the separated flow from the leading edge, the flaps (Fl) would suppress the flow separation from the upper edge, and the deflectors control the flow separation from the leading edge by streamlining the overall girder shape. Furthermore, horizontal plates (Hp) also play an important role of suppressing the flow separation from the leading edge but from the bottom of the section, the addition of a vertical plate (Vp) and a baffle plate would restrain the generation of vortices around the girder cross-section and the installation of spoilers could enhance the aerodynamic effects of the flaps.

The application of aerodynamic fairings and/or gratings to partially open up the deck floor has been widely accepted as possible stabilising methods against vortex-induced bridge vibration. However, these methods are usually applied after the wind tunnel investigation since their effectiveness depends upon the configuration of the primary structure and also the local wind conditions. Needless to say, aerodynamically stable girder shapes without any attachments should be sought and selected in advance. However, the bridge deck shapes are often chosen by considering only structural design or other different design factors such as economic efficiency. Consequentially, in many cases, aerodynamic countermeasures have been applied as the auxiliary methods by attaching devises on the original bridge section.

Buffeting

Significant bridge buffeting reduction by reducing the drag coefficient or increasing the stiffness is not easy for most cases. On the other hand, for the Yumemai Bridge [12], which is a movable floating bridge, aerodynamic appendages were applied for reducing the static wind loads. About 20% reduction of total wind load was achieved and it contributed to an economical design.

8.1.3.2 Structural Countermeasures

There are basically three ways for suppressing vibrations by structural means:

(a) Increase of girder mass;
(b) Increase of girder stiffness;
(c) Increase of structural damping.

The first method is for decreasing the relative contribution of the aerodynamic forces to the inertia force. The redundant weights such as concrete, for example, can be stuffed into the bridge girder for this purpose. The second method is for shifting the onset velocity by increasing the resonant frequencies. Temporary frames are sometimes used for this purpose during bridge erection. The third method is to increase the structure's capacity to dissipate vibration energy and thus reduce the dynamic response. Installation of artificial damping devices belongs to this method. This is most commonly adopted among the above listed methods.

The objective of installing the damping devices is mainly the suppression of vortex-induced vibration. The divergent vibrations such as galloping and flutter are rarely suppressed by the damping devices. A tuned mass damper (TMD) is one of the most common devices as artificial dampers. Installation of active control devices by applying sensors and actuators have not been practiced yet for actual bridges. At erection stages, the temporary scaffolds are sometimes built to increase their stiffness in addition to TMD.

Active control devices and modification of cable systems are being studied as a new development of the vibration control methods. Cable systems to increase stiffness, a passive control system using TPD (Tuned Pendulum Damper) [13] and temporary additional mass during violent storms [14, 15] are being examined for future design of super-long-span bridges, which are expected to have even longer spans than the Akashi-Kaikyo Bridge.

8.2 Examples of Countermeasures Against Wind-Induced Vibration

8.2.1 Aerodynamic Approaches

8.2.1.1 Akashi-Kaikyo Bridge

Case History of the Akashi-Kaikyo Bridge

This bridge across the Akashi-Kaikyo straits is the world's longest suspension bridge (3-span, 2-hinged truss-stiffened type) with a main span of 1,991 m and a total length of 3,911 m, and became open to the traffic on April 5, 1998. The Akashi-Kaikyo Bridge was originally designed as a suspension bridge with the center span of 1,780 m, serving both highway and railways. However, this design was altered to the present scheme due to a change of social circumstances in 1985.

Investigation into aerodynamics of the Akashi-Kaikyo Bridge as a highway bridge started in 1982. With regard to the aerodynamic stability, a wide variety of cross-sections [16] for the stiffening girder such as truss, box and their variations were investigated at first. It went through various comprehensive studies and the adoption of the truss-stiffened design at the end was based on two reasons: (1) it was deemed more appropriate in order to secure the aerodynamic stability; and

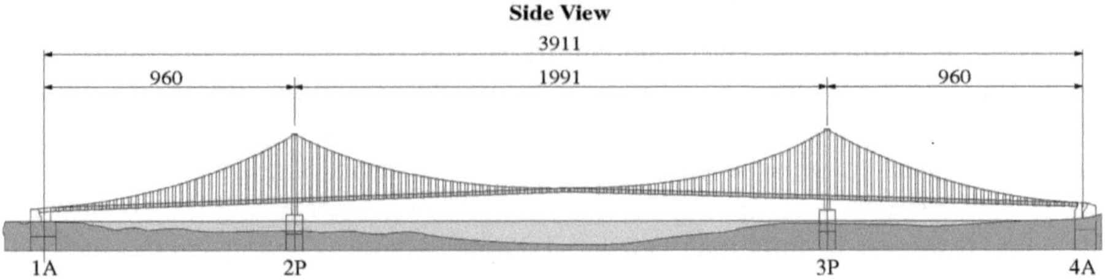

Fig. 8.2 General view of the Akashi-Kaikyo Bridge

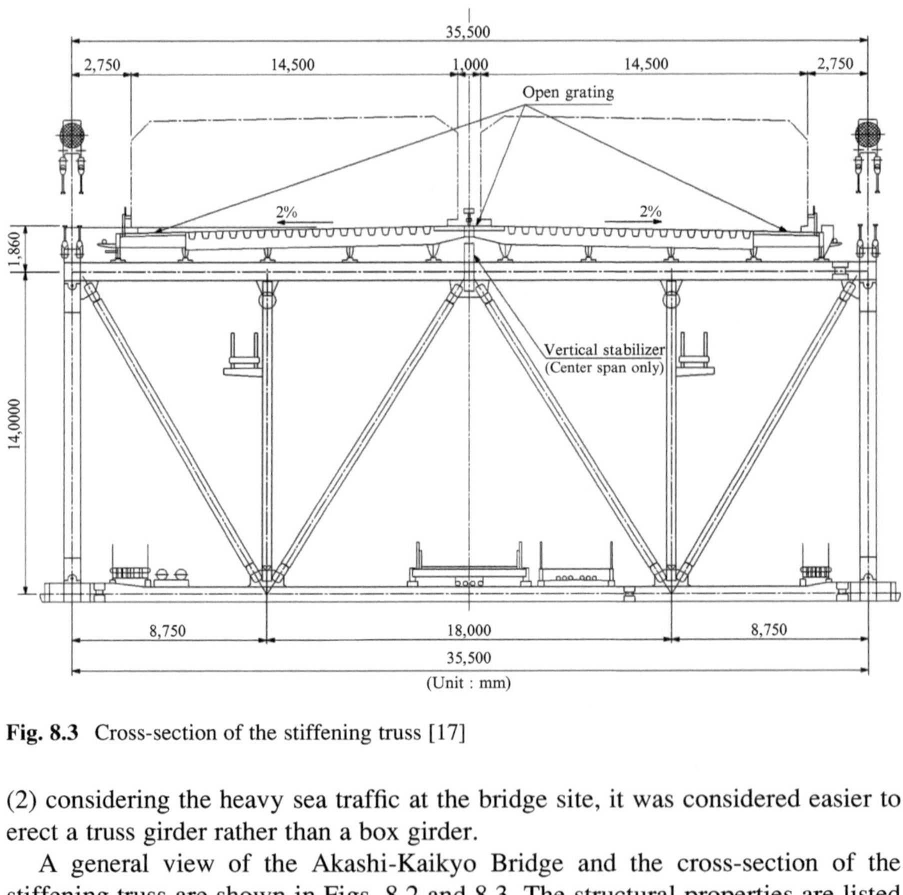

Fig. 8.3 Cross-section of the stiffening truss [17]

(2) considering the heavy sea traffic at the bridge site, it was considered easier to erect a truss girder rather than a box girder.

A general view of the Akashi-Kaikyo Bridge and the cross-section of the stiffening truss are shown in Figs. 8.2 and 8.3. The structural properties are listed in Table 8.6.

The main span of the Akashi-Kaikyo Bridge is much longer than the previous record holders such as the Humber Bridge with a main span of 1,410 m. Therefore, the wind resistant design for the Akashi-Kaikyo Bridge was considered to be more serious matter compared to the past experience with the other bridges. Figure 8.4 shows the reduced critical wind speed (*U*/*fB*, *U*: wind speed, *f*: frequency,

Table 8.6 Structural properties

Girder height (m)	14.0
Girder width (m)	35.5
Dead load (t/m)	
Girder	27.57
Cables	13.60
Total	41.17
Average (total span)	41.44
Natural frequency (Hz) 1st symmetric bending	0.064
1st symmetric torsion	0.135

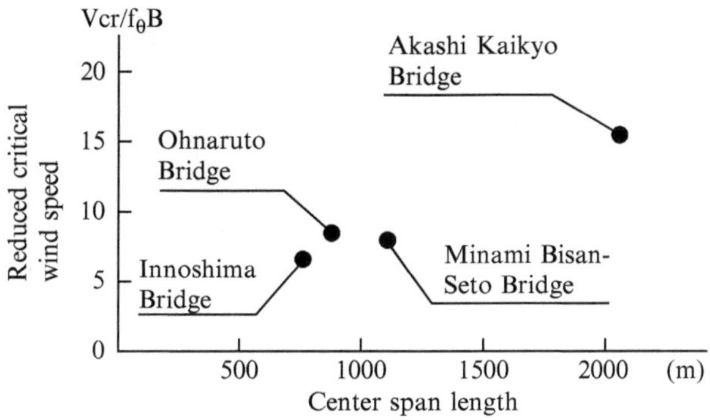

Fig. 8.4 Comparison of required level for wind resistant design [18]

B: deck width) of the Akashi-Kaikyo Bridge in comparison to some of the other Honshu- Shikoku bridges with 1 km-class center span.

As a result of the wind tunnel tests, the open gratings were provided along both shoulders ($B = 2.5$ m) as well as the median strip ($B = 3.5$ m) of the deck section for the entire bridge length in order to raise the onset wind speed of flutter. A vertical stabilizer was attached just beneath the open grating at the median strip, though the arrangement was limited only to the center span. Figure 8.5 shows the open grating along the shoulder, and Fig. 8.6 is the vertical stabilizer and the open grating along the median strip. An overview of the effects of open gratings and the vertical stabilizer on the aerodynamic characteristics of the Akashi-Kaikyo Bridge is discussed in the following sections.

Effect of Open Gratings

For the Truss-Stiffened Girder of the Highway-Railway Option

According to the wind tunnel tests which were conducted in 1977 using a 1:89 scale section truss girder model for the Akashi-Kaikyo Bridge of highway-railway combined use as shown in Fig. 8.7, flutter didn't occur up to 94 m/s when the deck had open gratings between traffic lanes. However, flutter was observed when the angle of attack of +3 to +5° was considered, once the gratings were closed. It was concluded that the open gratings should be provided on the deck in order to increase the flutter stability.

Fig. 8.5 Open grating along shoulders on the road deck (courtesy: Honshu-Shikoku Bridge Authority)

Fig. 8.6 Vertical stabilizer under the road deck and open grating along the median strip (courtesy: Honshu-Shikoku Bridge Authority)

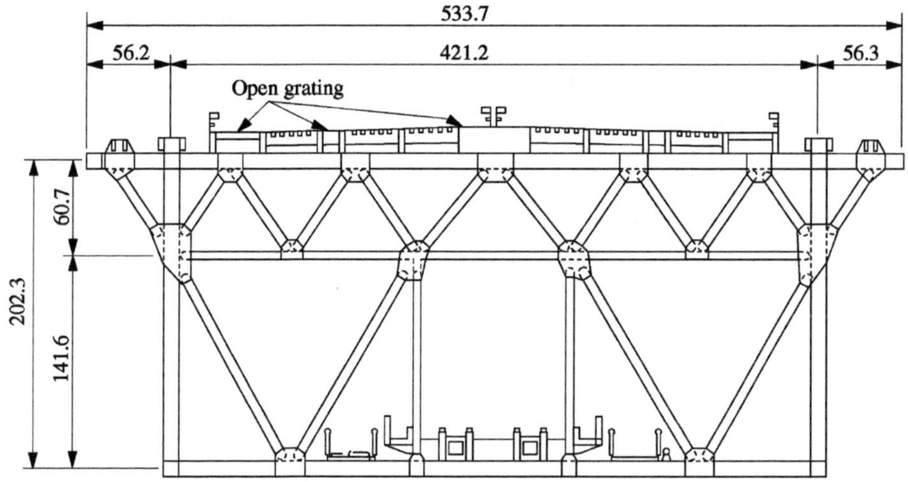

Fig. 8.7 Cross-section of Akashi Kaikyo Bridge model 1977 version (Model scale: 1/89)

For the Truss-Stiffened Girder of the Highway Option

Wind tunnel tests for the truss-stiffened girder of the highway bridge option, having a main span of 1,780 m, were conducted until 1984 to seek the following possibilities: (1) an extension of the center span from 1,780 m to 2,000 m, (2) possible removal of the open grating, (3) a reduction of the stiffening truss height: H, and (4) a reduction of the stiffening truss width. Figure 8.8 shows the test results studying the effects of grating between traffic lanes, and the conclusions are as follows.

- The flutter onset speed satisfied the code requirements of 78 m/s even though the utility lines such as electric power cables, telephone cables and water mains were arranged inside the truss. Flutter performance, however, became worse once the gratings between traffic lanes were closed.
- Even when the test conditions were set assuming the main span length of 2,000 m, the flutter characteristics satisfied the code requirements with stiffening girder type of case OT'-5*(H = 14 m). However, the flutter performance of case OT'-6*(H = 12 m) was found to be poor.

Effect of Vertical Stabilizer

A vertical thin box with a cross-section of 30 cm wide by 215 cm high installed at the center of the road deck was found to raise the flutter speed remarkably for both construction stage and after completion. Figure 8.9 shows the effect of the stabilizer based on the section model wind tunnel tests. The wind tunnel test with the aeroelastic full bridge model and the detailed flutter analysis confirmed that the vertical stabilizer was needed only for the center span.

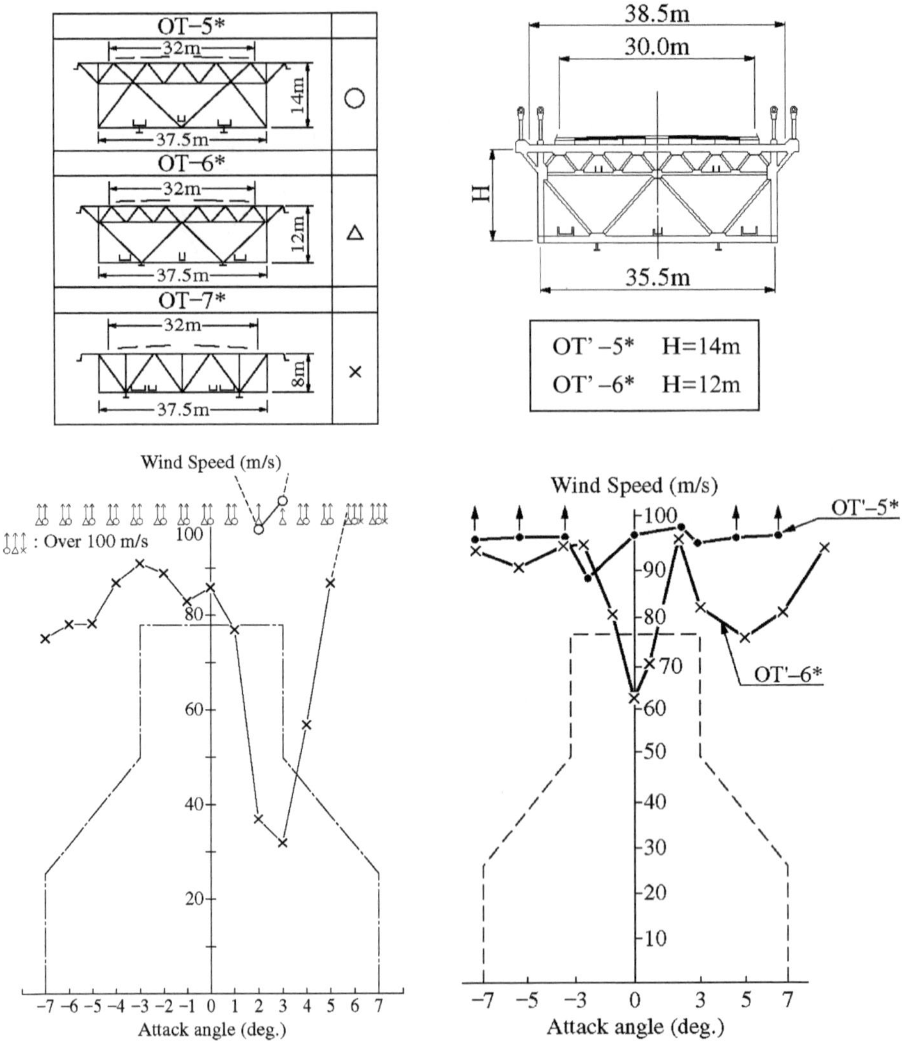

Fig.8.8 Comparison of different stiffening truss option cases (Section model test)

8.2.1.2 Tozaki Bridge

Case History of the Tozaki Bridge

The Tozaki Bridge, as shown in Fig. 8.10, consists of two multi-span box girders, and its maximum span length reaches 190.4 m. The deck width is 18.25 m and the deck depth varies from 4.5 to 8.2 m. The deck elevation is 40–50 m from the water level [19]. The bridge lies almost in parallel to the Tozaki peninsula, whose ridge elevation is nearly the same level as that of the bridge deck, which suggested a

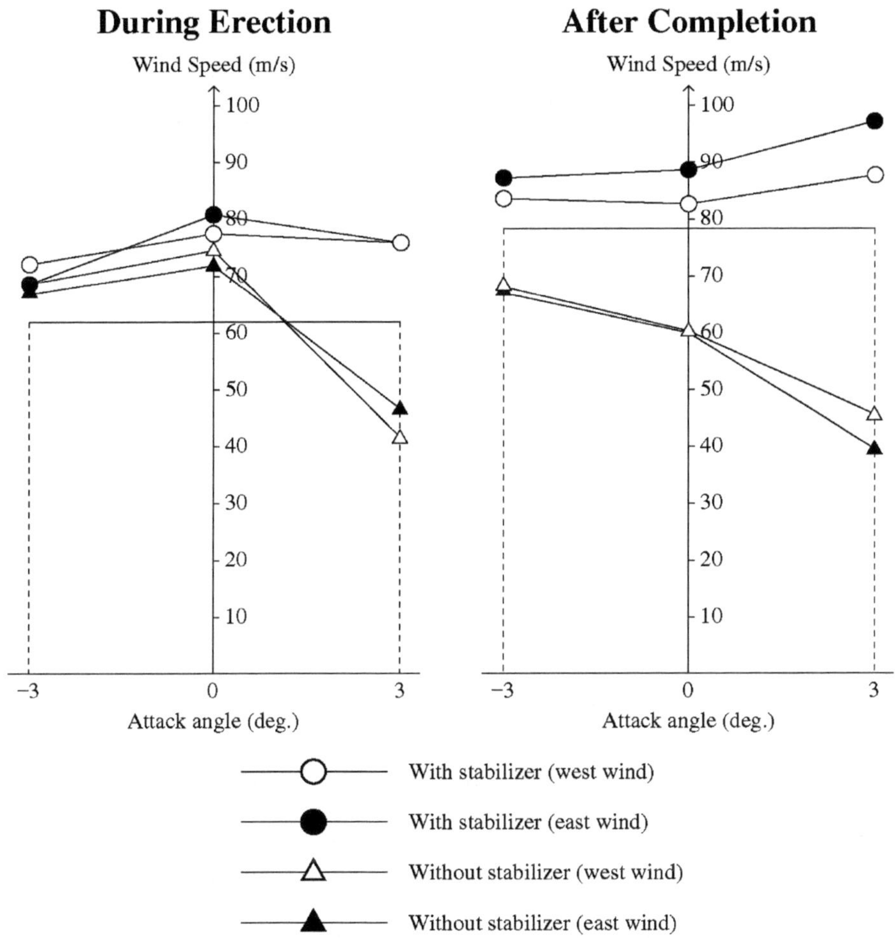

Fig. 8.9 Effect of vertical stabilizer (Section model test) [17]

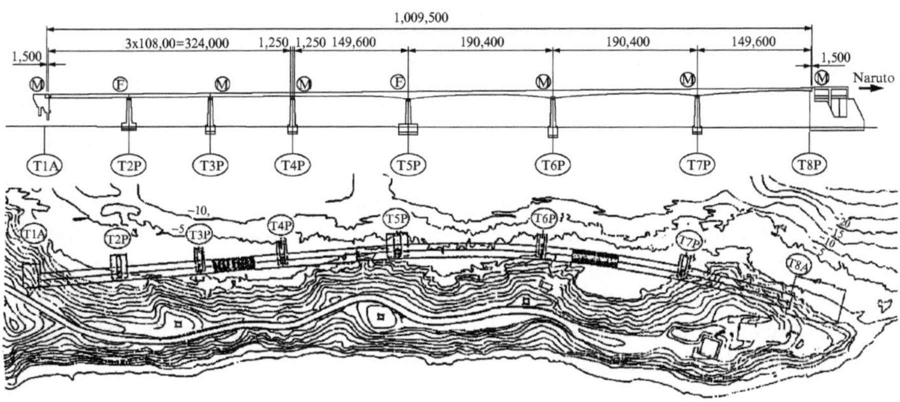

Fig. 8.10 General view of the Tozaki Bridge

Table 8.7 Structural conditions of the Tozaki Bridge (Protoype and wind tunnel test), Prototype: 4 span continuous steel box order bridge

		Prototype	Section model	Single span model	Full bridge model
Scale	1/n		1/50	1/80	1/120
Span		149.6 + 190.4 + 190.4 + 149.6 m	–	–	1,247 + 1,587 + 1,587 + 1,247 mm
Weight	W	14.58 tf/m	5.156 kgf/m	–	–
Width	B	18.25 m	365 mm	228 mm	152 mm
Depth	D	4.5–8.2 m	92,125.5 mm	56.25–102.5 mm	37.5–68.3 mm
Natural frequency (vertical) fj[Hz]	1st	0.399	2.918	3.692	5.58
	2nd	0.569	–	–	7.33
	3rd	0.808	–	–	9.86
	5th	0.991	–	–	11.20
Logarithmic decrement of structural damping δ_{Si}	1st	0.02[a]	0.02	0.02	0.02
	2nd–5th	0.02[a]	–	–	0.01

[a]Assumed value

possibility of incoming wind from the sea side with upward inclination. The deflection by the design live load reaches near the allowable limit as road bridges, which means that the structure is very flexible. The structural conditions of the bridge are as shown in Table 8.7. Because of these reasons, the bridge was carefully examined against vortex-induced vibrations as well as galloping by conducting wind tunnel tests and the means to stabilize the bridge without changing the basic configuration of the bridge.

Three kinds of wind tunnel tests, the 2-dimensional deck section tests, the single span model tests and the 3-dimensional full elastic model tests, were conducted. At the first stage of wind tunnel tests, a preliminary parametric study on aerodynamic stabilization was carried out by using the 2-dimensional deck section model. Next, influences of the surrounding topography, the effects of aerodynamic devices and the varying depth of the girder were investigated with a single span model. Finally, influence of the 3-dimensional topography of the peninsula and also of the curved bridge axis line on the aerodynamic behavior were investigated by a full bridge model with the proximity topography.

Through this series of wind tunnel tests, both a continuously placed double flap and the intermittently installed lower skirts were found to be effective for reducing the vortex-induced vibration and suppressing the galloping, respectively. A total weight of these attachments was 5,530 kN in all. Photos of the devices are given in Figs. 8.11 and 8.12. The deck section with the devices and the longitudinal arrangement of them are shown in Fig. 8.13.

Fig. 8.11 Double flap (Tozaki Bridge) (courtesy: Honshu-Shikoku Bridge Authority)

Fig. 8.12 Intermittently placed lower skirt and continuous double flap (Tozaki Bridge) (courtesy: Honshu-Shikoku Bridge Authority)

Wind Tunnel Tests

Design criteria were set as follows:

Allowable amplitude in the vortex-induced vibration: 200 mm,
Onset wind speed of galloping instability: 92 m/s or above.

The following conclusions are obtained in the series of wind tunnel tests. The response amplitude and wind speed are given in the full-scale values.

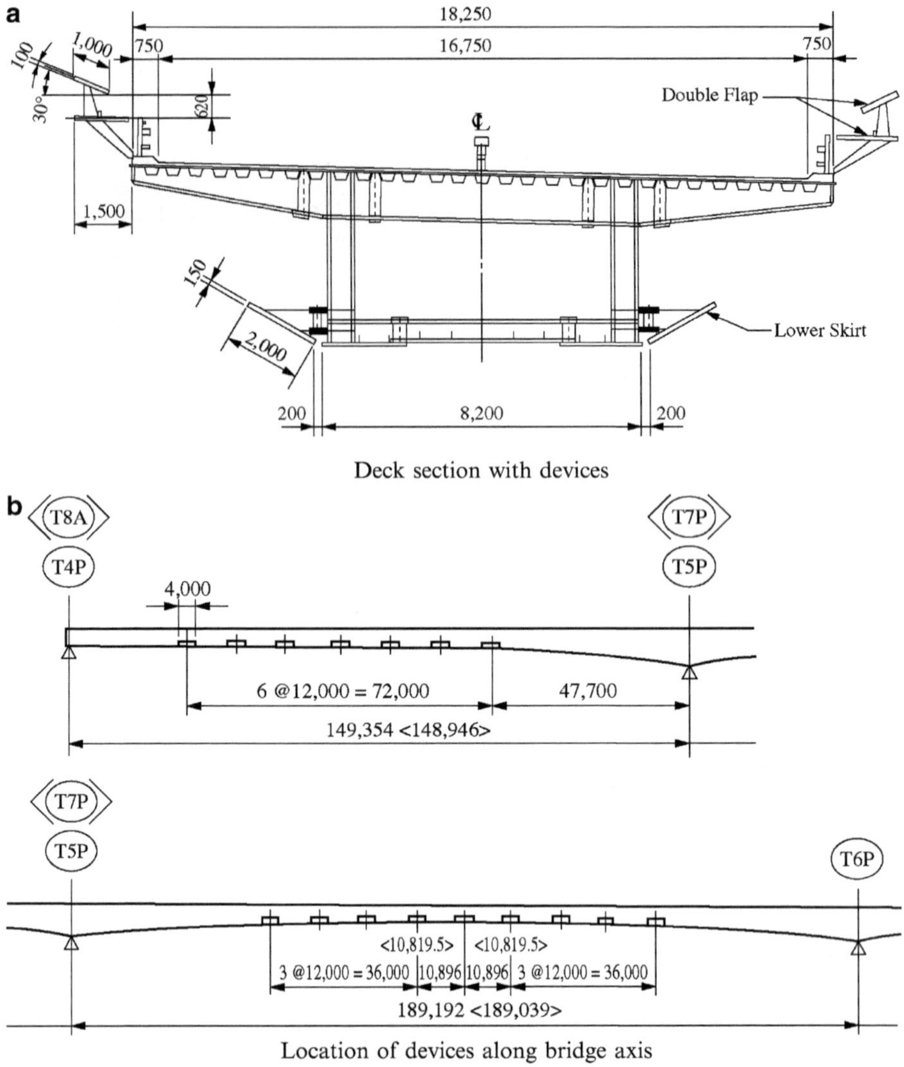

Fig. 8.13 Deck section with devices and location of devices in bridge axis

2-Dimensional Spring Supported Model Test

The deck sections at the *L*/2 point and at the *L*/6 point were examined (*L*: span length). The results of the *L*/2 (smaller *B*/*D*, and thus inferior in the stability) section without devices are summarized as;

- The vortex-induced vibration occurs at 15–20 m/s. The amplitude reaches 350 mm at the 0° angle of attack, 410 mm at 3°, 340 mm at 5°, and 680 mm at 10°.
- As for the galloping, the onset wind speed was 70 m/s at the attack angle of 0°, and yet only 30 m/s at 3–10°

Influence of modified corners of the box girder to suppress the galloping was investigated. Installation of the lower skirts was found to be effective, but this installation enlarged the amplitude of the vortex-induced vibration. However, the vortex-induced vibration was found to be mitigated by installation of the continuous double flaps outside the railings. Installation of the double flaps alone was found insufficient for suppressing the galloping, and it was thus concluded that the installation of both double flaps and intermittent lower skirts was adopted to be a good combination for securing aerodynamic stability.

3-Dimensional Full Aeroelastic Model Test

Tests were conducted both in smooth flow and turbulent flow, and the influence of the topography was also examined in smooth flow. In smooth flow test, to which the aerodynamic devices were not placed, amplitude of vortex-induced vibration of the 1st mode reached 500 and 1,100 mm corresponding to the case without and with the topographic model, respectively. Galloping was observed at about 35 m/s, with or without the topographic model.

When the aerodynamic devices were installed to the 3-dimensional bridge model, neither vortex-induced vibration nor galloping was observed without terrain model in place. Even with the terrain model, vortex-induced vibration was decreased to 250 mm and no galloping was observed.

The influence of topography was also examined in turbulent flow which had the along-wind turbulence intensity of 9–14% and the horizontal scale of turbulence of about 60 m. Vortex-induced vibration did not occur when the aerodynamic devices were provided, but the amplitude of the vortex-induced vibration reached 400 mm when the devices were not arranged. The responses of the 3-dimensional full aeroelastic model tests are summarized as shown in Fig. 8.14.

Field Observation of the Tozaki Bridge

Structural logarithmic decrement measured on the actual bridge exceeded 0.1, which was about five times greater compared to the assumed value for the wind tunnel tests. The maximum wind speed experienced at the site so far is 29 m/s and no obvious bridge vibration has been reported.

8.2.2 Wind-Induced Vibration and Control of Tokyo Bay Aqua-Line Bridge Girder

In most of long-span bridges, wind tunnel testing is performed prior to the bridge's final design and consequently various control means are implemented whenever necessary. Hence, it is rather rare to observe noticeable wind-induced vibration in

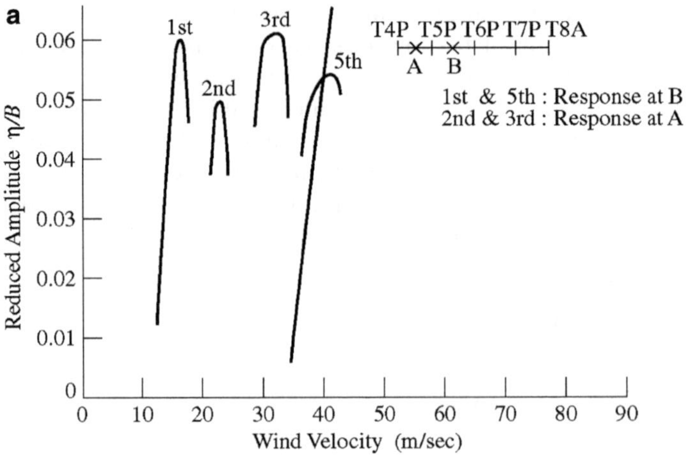

Basic section (w/o devices) in smooth flow

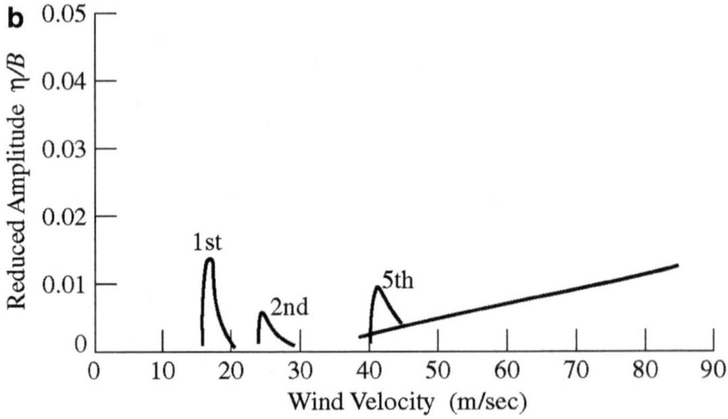

Improved section (with double flap & intermittent skirt) in smooth flow

Fig. 8.14 3D-full model test (with the topographic model)

actual bridges. If such vibrations should indeed occur in existing bridges, this would provide excellent opportunities to study *real* fluid-structure interaction and to clarify the validity of the wind tunnel tests. In case of the Great Belt suspension bridge, the Rio-Niteroi steel box-girder bridge and so on, it is reported that vortex-induced vibration was actually observed and extremely important findings are expected to result from these examples [20–27].

The Tokyo Bay Aqua-Line is a combined tunnel and multiple bridge route that includes a ten-span continuous steel box-girder bridge, as shown in Figs. 8.15 and 8.16. In the design phase, various wind tunnel tests revealed that vortex-induced vertical flexural vibration would develop in the bridge under the wind normal to the bridge axis. Although a variety of aerodynamic mitigation strategies were tested, complete suppression of the vibration was not achieved.

Fig. 8.15 Tokyo Bay Aqua-Line Bridge

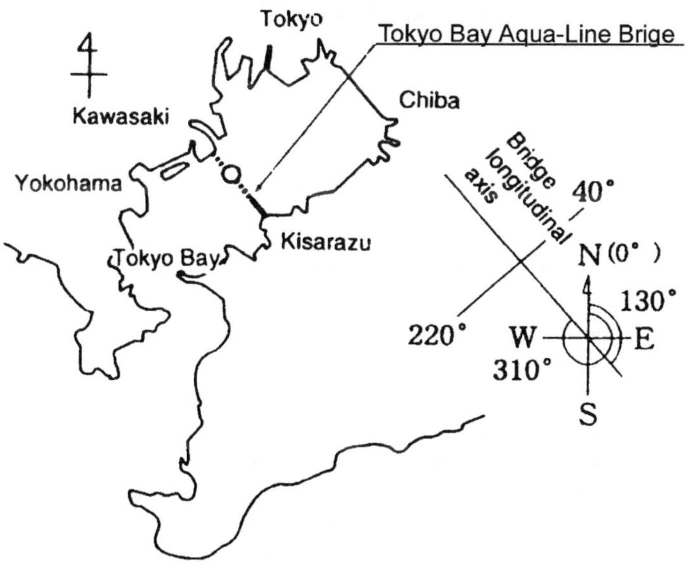

Fig. 8.16 Bridge location map

This bridge also provided access for construction of the man-made reclaimed island and the shield tunnel of the crossing; the highway was not put into service until long after completion of the bridge. Consequently, a decision was made not to install TMDs to reduce the vibration unless excessive vibration was observed during monitoring of the girder behavior.

Fig. 8.17 Video photos of vortex-induced first-mode vibration of the bridge

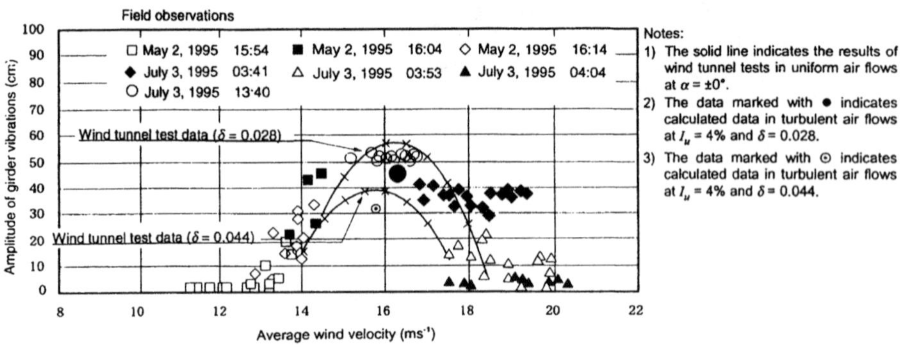

Fig. 8.18 Comparison of amplitudes of girder vibrations

8.2.2.1 Vortex-Induced Vibration Observed in Bridge

Girder erection was completed in October 1994. Significant vibration was observed under prevailing winds from the southwest direction during the following December, and thereafter vibration, probably of a vortex-induced in nature, was observed under certain conditions. This vortex-induced first-mode vibration peaked at a wind speed of 16–17 m/s, with a maximum amplitude exceeding 50 cm. Figure 8.17 shows the frozen video images of the bridge deck during the vortex-induced vibration at two instants of time corresponding to the maximal upward and downward displacements. Comparing two photos, one can easily understand the large amplitude of the girder vibration.

Figure 8.18 presents displacement amplitudes under strong wind almost normal to the bridge axis. These observations were made over a period of 30 min on May 2, 1995, and over 40 min on July 3, 1995. The data plotted in this figure are values averaged over a period of 30 s. Two sets of data with large response amplitudes, at 16:04–16:14 on May 2 and 13:40–13:50 on July 3, 1995, were subjected to frequency analysis, which indicated that the first-mode vibration dominated the response. The displacement amplitude of the girder in Fig. 8.18 was estimated from the measured acceleration and the natural frequency of the first mode of vibration assuming the following relationship:

$$\eta = \alpha/(2\pi f_1)^2 \tag{8.1}$$

where η = amplitude of vibration (cm); α = measured maximum acceleration (cm/s^2); and f_1 = the first mode natural frequency of the girder (=0.34 Hz).

The maximum acceleration recorded was 253 cm/s^2, and the amplitude calculated from (8.1) was about 54 cm. In this instance, the wind direction was southwesterly at 205° from the north; i.e., 15° to the east of the bridge transverse axis, and the turbulence intensity I_u in the primary wind direction was between 4 and 6%.

As shown in Fig. 8.18, the vortex-induced vibration of the first flexural mode occurred at the wind speed ranging from 13 to 18 m/s, and the maximum amplitude was reached at 16–17 m/s. The solid curves in Fig. 8.18 represent data measured from the wind tunnel tests of a ten-span full bridge model in smooth flow and, for this comparison, they are adjusted to logarithmic decrement of δ = 0.028–0.044, which was obtained from the forced vibration test of the actual bridge. It can be seen from this figure that the field observations agree well with the wind tunnel test data of the three-dimensional full model of the as-built design at a scale of 1:170 in uniform air flow at the wind angle of attack, $\alpha = 0°$. The reason for making the comparison in uniform flow is that most of the wind tunnel tests were carried out in uniform flow so as to obtain conservative results.

8.2.2.2 Tuned Mass Damper

The tuned mass dampers (TMDs) for this bridge were designed to be accommodated inside the box girder. They have a structure comprising major and minor frames in the shape of a pantograph (Fig. 8.19a). The TMD weight is mounted on the lower end of the major frame and moves up and down with girder vibrations. The coil springs and oil dampers act as the minor frame moves up and down. This design makes it possible to considerably reduce the size of TMDs. Eight TMDs for the first mode and eight TMDs for the second mode were installed (Fig. 8.19b).

The TMDs for controlling the first- and second-mode vibration were brought into operation beginning in, respectively, January and late July 1996. Figure 8.20 compares field observations made before and after installation of the TMDs for the first mode. In this figure, vibration records under similar wind conditions, in both speed and direction, before and after installation of the TMDs were taken over 10 min.

As can be seen in the comparison, the maximum acceleration and displacement amplitudes of the girder before installation of the TMDs were 191 gal. and 40.7 cm

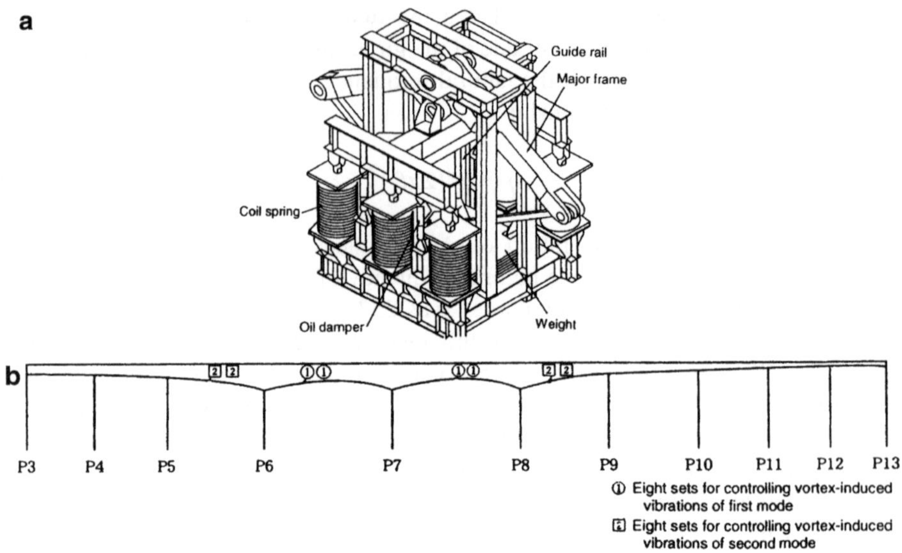

Fig. 8.19 Newly developed tuned mass damper (TMD) and location of tuned mass dampers in girder

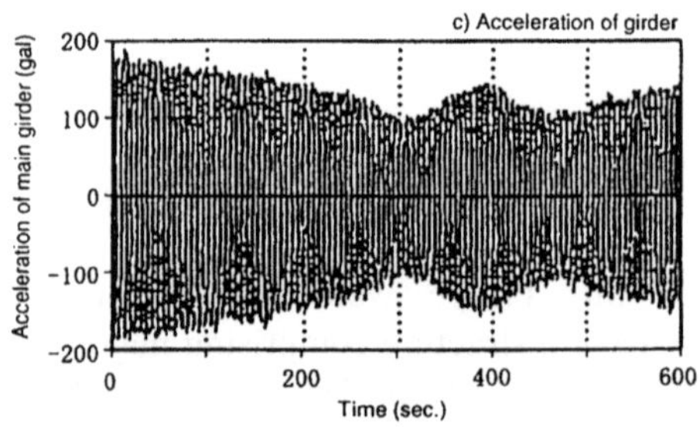

	13:50-14:00 on Nov. 1, 1995	19:02-19:12 on Mar. 11, 1996
Mean wind velocity	17.8 m/s	17.5 m/s
Average wind direction	221°	220°
Turbulence intensity	5.5%	7.7%
Max. acceleration of girder	191 gal	25.4 gal
Max. half amplitude of TMDs		164 mm

(1) Before installation of TMDs
13:50-14:00 on November 1, 1995

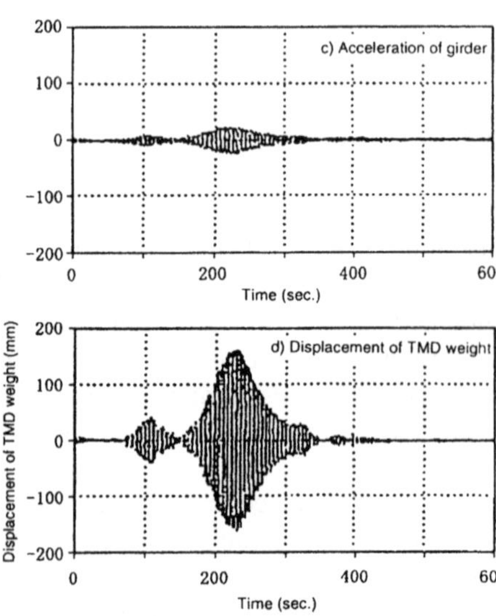

(2) After installation of TMDs for controlling first mode vibrations
19:02-19:12 on March 11, 1996

Fig. 8.20 Effects of tuned mass dampers (TMDs) on controlling first-mode vibrations (*1*), (2) (as converted using the first-mode natural frequency), whereas after installation of the TMDs they were noticeably reduced to 25.4 gal. and 5.4 cm, respectively. The maximum displacement amplitude of the TMD weights was 164 mm.

A point worth noting is that the installation of vertical plates outside the railings was nearly completed on November 1, 1995, which were found to reduce the 3rd and 4th mode vibration.

References

1. Japan Road Association (1991) Wind resistant design manual for highway bridges (in Japanese). Maruzen, Tokyo. Also revised version in 2007
2. Shiraishi N, Matsumoto M (1982) On vortex-induced oscillations of bluff cross sections used for bridge structures (in Japanese). In: Proceedings of the JSCE (322):37–50
3. Takeda K, Sonobe Y, Hashimoto M (1986) Experimental study to estimate the amplitude of vortex-induced vibration (in Japanese). In: Proceedings of the 9th National Symposium on Wind Engineering, Tokyo, pp 265–270
4. Kawakami Y (1994) The Skygate Bridge (in Japanese). Doboku-Seko 35(10):41–48
5. Japanese Society of Steel Construction (1997) Wind resistant engineering of structures (in Japanese). Tokyo Denki University Press, Tokyo, Japan, pp 363–372
6. Yamada H (1995) Approach to wind resistant engineering, Kensetsu-Tosho, Tokyo (in Japanese)
7. Japan Society of Civil Engineers (1985) Vibration handbook for civil engineers, JSCE, Tokyo (in Japanese)
8. Bridge and Wind Editorial Group (1994) Bridge and wind, Kyoto (in Japanese)
9. Kobayashi H, Nagaoka H (1992) Active control of flutter of a suspension bridge. J Wind Eng Ind Aerod 41(1–3):143–151
10. Wilde K, Omenzetter P, Fujino Y (2001) Suppression of bridge flutter by active deck-flaps control system. J Eng Mech ASCE 127(1):80–89
11. Hiejima S (2005) Feedback control of vortex shedding around a bluff body by velocity excitation. Int J Comput Fluid Dynam 19(1):87–92
12. Maruyama T et al (1998) Plan and design of Yumeshima-Maishima Bridge (in Japanese). Bridge and Foundation Eng 32(2):15–24
13. Okada T, Honke K, Sugii K, Shimada S, Kobayashi H (1998) Suppression of coupled flutter of a bridge deck by tuned pendulum damper. In: Proceedings of the 2nd international conference on structural control (2WCSC) 3:1747–1756
14. Nakazaki S, Yamaguchi H (1998) Very long suspension bridges using the temporary mass method. IABSE Reports, (IABSE Symposium Kobe 1998) 79:415–420
15. Nakazaki S, Yamaguchi H (1999) Preliminary design of very long suspension bridges using the temporary additional mass method against a storm. J Wind Eng Ind Aerod 83:317–326
16. Miyata T, Tada K, Katsuchi H (1993) Wind resistant design considerations for the Akashi Kaikyo Bridge. In: Proceedings of the international seminar on utilization of large boundary layer wind tunnel, Tsukuba, pp 79–100
17. Honshu-Shikoku Bridge Authority (1998) The Akashi-Kaikyo bridge: design and construction of the world's longest bridge, HSBA, Kobe
18. Miyata T, Sato H, Kitagawa M (1993) Design considerations for superstructures of the Akashi Kaikyo bridge. In: Proceedings of the international seminar on utilization of large boundary layer wind tunnel, Tsukuba, pp 121–139

19. Honshu Shikoku Bridge Authority (1983) The First construction bureau Kobe: construction of Tozaki bridge -Largest prefabricated block erection of steel box girders were executed successfully, HSBA, Kobe, Japan (in Japanese)
20. Hirai S, Honda A, Kato H, Yoshida O, Okauchi I (1993) Aerodynamic stability of Trans-Tokyo bay highway bridge. J Wind Eng Ind Aerod 49:487–496
21. Yoshida Y, Tokita H, Fujino Y, Katsuura K (1997) Excitation test on a box girder bridge with continuous steel deck and multiple spans and its vibration characteristics (in Japanese). J Struct Eng JSCE 43A:725–736
22. Yoshida Y, Fujino Y, Tokita H, Honda A (1999) Wind tunnel study and field measurement of vortex-induced vibration of a continuous steel box girder in Trans-Tokyo Bay Highway (in Japanese). J Struct Mech Earthquake Eng JSCE 633/I-49:103–117
23. Yoshida Y, Fujino Y, Satom H, Tokita H, Miura S (1999) Control of vortex-induced vibration of a continuous steel box girder in Trans-Tokyo Bay Highway (in Japanese). J Struct Mech Earthquake Eng JSCE 633/I-49:119–134
24. Fujino Y, Yoshida Y (2002) Wind-induced vibration and control of Trans-Tokyo bay crossing bridge. J Struct Eng ASCE 128(8):1012–1025
25. Larsen A, Esdahl S, Andersen JE, Vejrum T (1999) Vortex shedding excitation of the Great Belt suspension bridge. In: Larsen A, Larose GL, Livesey FM (eds) Wind engineering into the 21st century. Balkema, Rotterdam, pp 947–954
26. Battista RC, Pfeil MS (2000) Reduction of vortex-induced oscillations of Rio-Niteroi bridge by dynamic control devices. J Wind Eng Ind Aerod 84:273–288
27. Frandsen JB (2001) Simultaneous pressures and accelerations measured full-scale on the Great Belt East suspension bridge. J Wind Eng Ind Aerod 89:95–129

Chapter 9
Vibration Control for Bridge Towers and Field Measurement

9.1 General

9.1.1 Introduction

In Japan, steel towers are widely employed for the cable-supported bridges mainly due to high seismic design forces. Steel towers of suspension and cable-stayed bridges are very flexible and have generally low natural frequencies. In addition, the structural damping of towers tends to be low. Furthermore, tower shafts have usually rectangular or other aerodynamically bluff cross-sections. As a result, they are extremely susceptible to wind effects, particularly at their erection stage, and would suffer from wind-induced vibration such as vortex-induced vibration and/or galloping if no countermeasures are taken.

Once the bridge is completed, towers of suspension and cable-stayed bridges are less problematic regarding the wind-induced vibrations because the towers are usually constrained in direction of the bridge axis by cables and have large stiffness in the direction normal to it. However, the towers are prone to the adverse wind effects during erection because they are free standing (cantilever) structure then and often exposed to wind in an open terrain with low turbulence intensity.

It has been well known that in 1964, during its construction, the towers of the Forth Bridge in U.K. vibrated with the maximum amplitude of 1.1 m in 9–10 m/s wind. In order to suppress the vibration, a heavy sliding block was attached to the top of the free standing tower by cables [1]. In 1971, the sliding block method was applied to the Kanmon Bridge for the first time in Japan (Fig. 9.2). Kanmon is a suspension bridge with the center span of 712 m, the longest span in Japan at that time. In recent years, newly developed countermeasures, such as tuned mass dampers and other active dampers, have been adopted to the towers of major suspension and cable-stayed bridges at the construction stage.

Along with the development of large cable-stayed bridges in Japan, research on the vibration control technology also made progress and some of them have been applied to the bridge towers in reality. In the case of the Katsushika-Harp Bridge,

Y. Fujino et al., *Wind Resistant Design of Bridges in Japan*,
DOI 10.1007/978-4-431-54046-5_9, © Springer 2012

a cable-stayed bridge with single column towers, the aerodynamic deflectors of a curved shape were attached to each corner of the rectangular tower shafts to suppress both the vortex excitation and galloping which were predicted to occur after completion as well as during the construction. In the case of the Meiko Higashi (East) Bridge, a cable-stayed bridge with A-shaped towers, the location of a horizontal strut was found to aerodynamically affect in-plane divergent vibration caused by wind in the longitudinal direction and it was determined by a wind tunnel test.

As mentioned earlier, it was indicated by wind tunnel tests and also by analyses that the unacceptable wind-induced vibration could take place at the towers of the Akashi-Kaikyo Bridge (1998) and the Kurushima-Kaikyo Bridges (1999) after their completion. In the Akashi-Kaikyo Bridge, the tuned mass dampers (TMDs) were installed inside the tower shafts to control the vibration. In the case of the Kurushima-Kaikyo Bridges, the amplitude of the vibration was taken into account at the structural design of the towers by using thicker plates.

More recently, various types of active vibration dampers have been contrived and applied to the towers of suspension bridges and cable-stayed bridges in Japan. A hybrid mass damper was, as the first time, introduced to the Rainbow Bridge in Tokyo in 1991.

9.1.2 Basic Procedure to Counter Wind-Induced Vibration

A standard procedure of examining the aerodynamic stability and its control, if necessary, is as follows:

9.1.2.1 Determination of Acceptable Amplitude

Acceptable limit of vortex-induced vibration amplitude is, first of all, determined as a part of the basic examination of aerodynamic stability of towers. Below the design wind speed, galloping must be avoided at any cost, since it could be destructive to the structure.

Safety of Members and Facilities

The acceptable limit of dynamic amplitude is determined by conditions not to have failure and also by considering fatigue damage of tower members and construction facilities such as cranes.

Workability During Construction

The acceptable limit of amplitude is also restricted by the acceptable acceleration of the vibration for workers to maintain workable conditions. In general, 50 gal is adopted as the critical value of the acceleration.

9.1.2.2 Examination of Aerodynamic Stability

Vibration Analysis

Prior to the wind tunnel tests, vibration analyses are conducted to obtain the natural frequencies and corresponding vibration modes of the towers at each of the typical construction stages.

Wind Tunnel Tests

Wind tunnel tests are carried out with models considering various construction stages and, if necessary, also of the completion stage. For the damping ratio of the structure at the construction stage, a lower value is used than that employed for the completed structure; the damping ratio was determined by considering the results of field observation carried out on existing bridges.

9.1.2.3 Selection of Methods to Control Wind-Induced Vibrations

Based on the wind tunnel tests, countermeasures to suppress wind-induced vibrations are selected.

In general, aerodynamic means is studied first considering the tower configuration and cross-sections of the shafts. The purpose of this is to minimize the vibration amplitude so that the required addition of damping, if it is required, would be minimized, too. Then, the mechanical countermeasures are studied to provide enough damping to suppress the vibration down to the acceptable limit as mentioned above.

The type of vibration control should be determined in accordance with the other design conditions, frequencies and required damping, and the condition of the construction site.

9.1.3 Methods to Control Vibrations

Countermeasures to suppress vibration are classified into two groups; the aerodynamic means and the mechanical methods.

9.1.3.1 Aerodynamic Means

Aerodynamic means are classified basically into two groups: (a) changing the basic cross-section shapes, such as cutting off the corners or making slits etc.; and (b) attaching the aerodynamic appendages, such as deflectors, side-plates etc., as shown in Fig. 9.1.

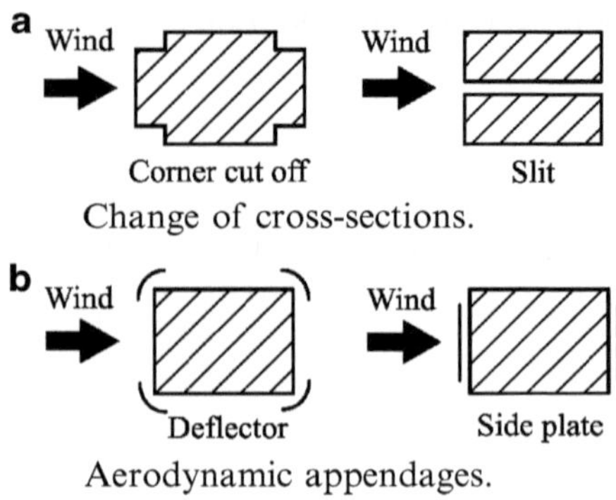

Fig. 9.1 Examples of aerodynamic countermeasures [2]. (**a**) Change of cross-sections. (**b**) Aerodynamic appendages

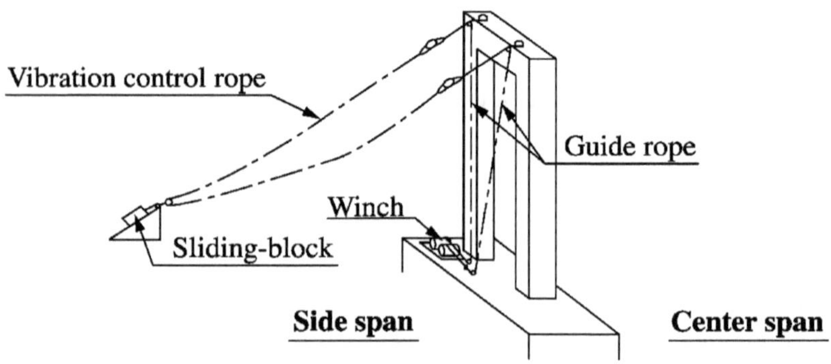

Fig. 9.2 Example of vibration control (Sliding block method)

Making a square cut at each corner of a rectangle cross-section is known to be effective in preventing the vortex-induced vibration. However, it is also known that its effectiveness depends on the Reynolds number. Wind tunnel tests for examining the aerodynamics of the square-corner-cut sections should be conducted keeping the Reynolds number effects in mind [3, 4] (see also Sect. 6.4.1).

The addition of aerodynamic appendages must be selected by considering not only their effectiveness but also the maintenance, durability and aesthetics of them.

9.1.3.2 Mechanical Control of Vibrations

In the 1980s, the sliding-block method as mentioned earlier, and also a combination of oil-dampers and attachment of weights (Fig. 9.2) were employed as mechanical means to control wind-induced vibrations.

Tuning-type countermeasures such as tuned mass damper (TMD), tuned liquid damper (TLD), tuned sloshing damper (TSD) and tuned liquid column damper (TLCD) were commonly used in recent bridges (Fig. 9.3). Active-type

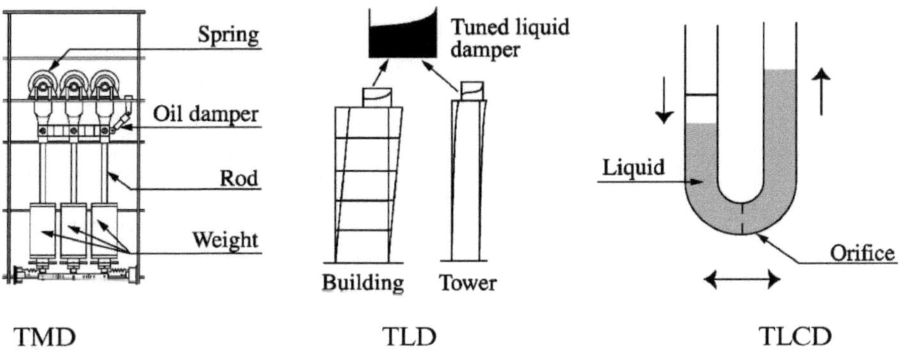

Fig. 9.3 Example of vibration dampers (tuned vibration control) [5]

countermeasures such as active mass damper (AMD) and hybrid mass damper (HMD) were lately adopted because they are more effective to gain required damping compared with the conventional tuning-types mentioned above.

9.2 Examples of Vibration Suppression and Field Measurement

9.2.1 Akashi-Kaikyo Bridge

9.2.1.1 General Procedure

The towers of the Akashi-Kaikyo Bridge were more flexible compared with those of already existing bridges then, because of its world record height of 297 m above sea level. Thus, the vibration control was an important issue not only at its construction stage but also after completion.

Wind tunnel tests were conducted to obtain the wind speed at which vortex-induced vibration occurs and the relationship between the Scruton number and the amplitude of the vibration. The test results were compared with the allowable amplitude obtained from structural analyses, and countermeasures to suppress vibration were examined [6, 7].

Because the countermeasures were required not only in the construction stage but also after completion in the case of the Akashi-Kaikyo Bridge, vibration control devices were designed so that they could be used both after completion and during the construction stage.

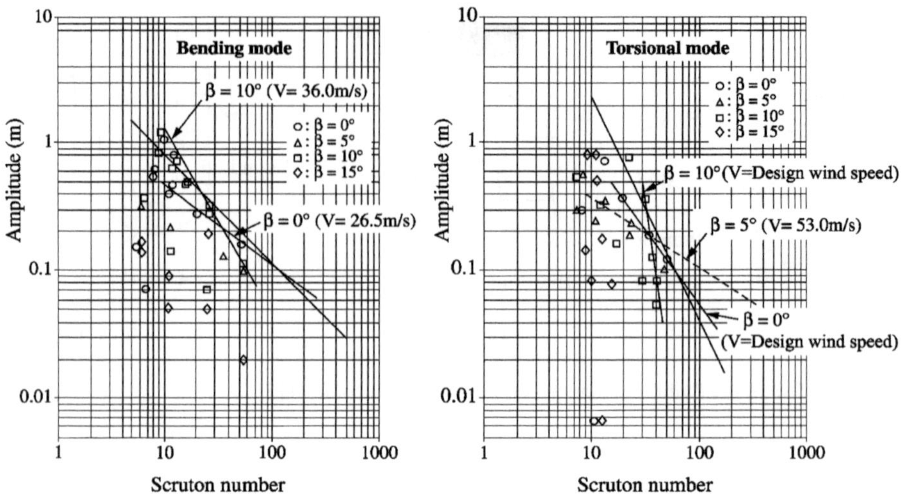

Fig. 9.4 Maximum amplitude vs. Scruton number (after completion) obtained by wind tunnel test using an elastic model (β: wind direction) [7]

9.2.1.2 Wind Tunnel Test Results

Figure 9.4 shows the relationship between the Scruton number and maximum vibration amplitude obtained from the wind tunnel tests conducted for the completion stage of the suspension bridge. Damping required for the vibration control devices was determined based on these relationships.

9.2.1.3 Examination of the Vibration Control Methods

Examination of Tower Configuration

The configuration of the tower bracings and the cross-section of the tower shafts were determined by the wind tunnel tests so that the vortex-induced vibration and galloping instability would be avoided as much as possible. Though a portal type tower was deemed aesthetically favourable, a truss type was adopted at the end because of its better aerodynamic performances. A square-corner-cut cross-section was employed for the tower shafts because of its effectiveness in reducing the amplitude of vortex-induced vibration, as shown in Fig. 9.5. As mentioned earlier, this section was found to possess Reynolds number dependency [3, 4].

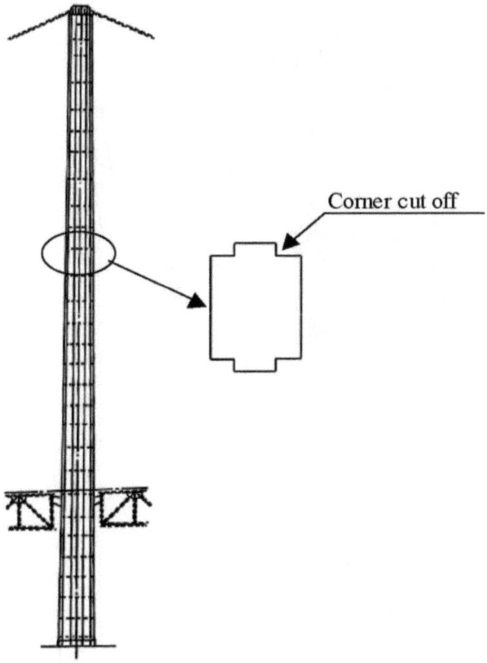

Fig. 9.5 Cross-sectional shape with corner cuts [4]

Countermeasures Using Damping Devices

Even after the shaft cross-section was optimized as above, the induced amplitude after completion was still higher than acceptable limit, which was determined based on the corresponding stress level. The use of damping device, the tuned mass dampers shown in Fig. 9.6 for this case, was decided as a result. As shown in Fig. 9.7, 8 sets of TMDs for the first bending mode and 12 sets of TMDs for the first torsional mode were installed in the tower shafts. In order to meet the "fail-safe" concept, in addition to these TMDs, some oil dampers were installed between the tower shafts and the stiffening truss girders. It was designed in such a way that even the TMDs or the oil dampers alone would be sufficient for suppressing the motion down to the acceptable level.

For the construction period, wind tunnel tests were conducted on the tower predominant vibration modes for which vibration control was deemed necessary. Based on the wind tunnel test results, the Hybrid Mass Dampers (HMDs, Fig. 9.6b), which actively control the weights with a wide frequency range during construction, were used. In addition, the TMDs for the completion stage were utilized in the erection stage by adjusting their dynamic characteristics.

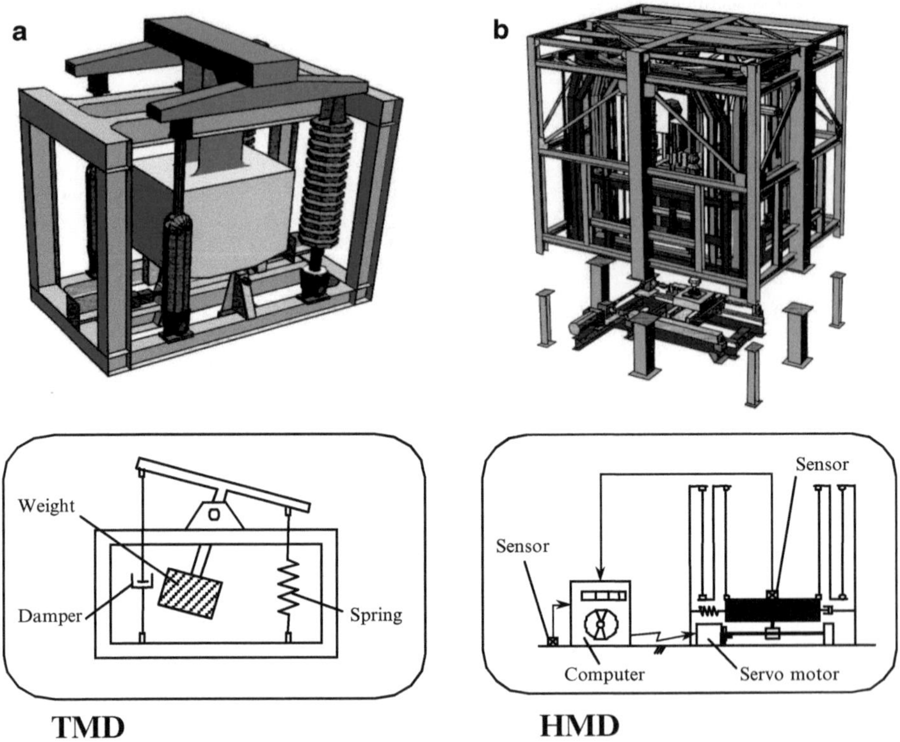

Fig. 9.6 Vibration control devices used [6]. (**a**) *TMD*, (**b**) *HMD*

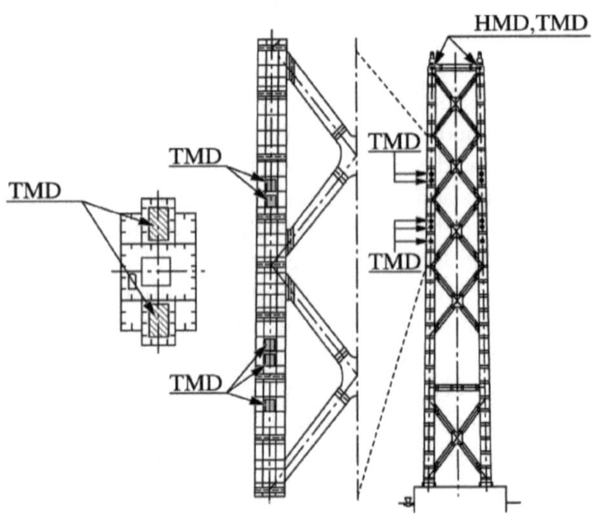

Fig. 9.7 Placement of control devices in the tower [6]

Table 9.1 Natural frequency and logarithmic decrement

Vibration mode (natural frequency)	Test case	Damping device				Logarithmic decrement		Maximum amp. (cm)
		HMD[a]	TMD1	TMD2	TMD3	Measured	Design	
1st out-of-plane vibration mode (0.126 Hz)	1	×	×	×	×	0.0067	0.0100	62
	2	P	×	×	×	0.028	0.0244	28
	3	A	×	×	×	0.105	0.0752	19
	4	P	○	○	○	0.038	0.0244	21
	5	A	○	○	○	0.111	0.0752	21
2nd out-of-plane vibration mode (0.673 Hz)	6	×	×	×	×	0.038	0.0100	1.3
	7	×	×	○	×	0.080	0.0454	0.9
	8	P	○	○	○	0.096	0.0454	1.0
Torsional vibration mode (0.471 Hz)	9	×	×	×	×	0.028	0.0100	5.3
	10	×	○	×	○	0.075	0.0418	3.0
	11	P	○	○	○	0.075	0.0418	3.0

[a]*P* Passive, *A* Active, ○ in use, × not in use

9.2.1.4 Vibration Test

After the completion of the tower, a series of vibration tests were conducted with HMDs, which were placed at the top of the tower and gave excitation. Table 9.1 shows the frequencies and damping obtained from the tests. Damping was measured when HMDs were set both in active and passive modes, and the TMDs were either active or non-active. The frequencies obtained from the tests agreed well with the analytical results. The logarithmic decrement for the first bending mode was 0.0067, which was less than 0.01 assumed for the wind tunnel tests. The damping values for the other vibration modes were found higher than 0.01, which were likely to be influenced by the existence of the tower cranes.

9.2.1.5 Field Measurements During the Tower Construction

When the tower was under construction, there were some occasions to measure wind-induced response of the tower and also the performance of vibration control devices. Figure 9.8 shows the observed relationship, together with the wind tunnel test results, between wind speed and the amplitude of the tower shaft vibration when the vibration control devices were effective. Major findings are as follows:

(a) Observed data showed generally good agreement with the wind tunnel test results. However, in the field observation, the wind speeds at which the vibration was observed did not change so much for different wind directions as they did in the wind tunnel cases. This would be attributed to the fact that both wind speed and its direction were always fluctuating in reality compared to the laboratory tests.

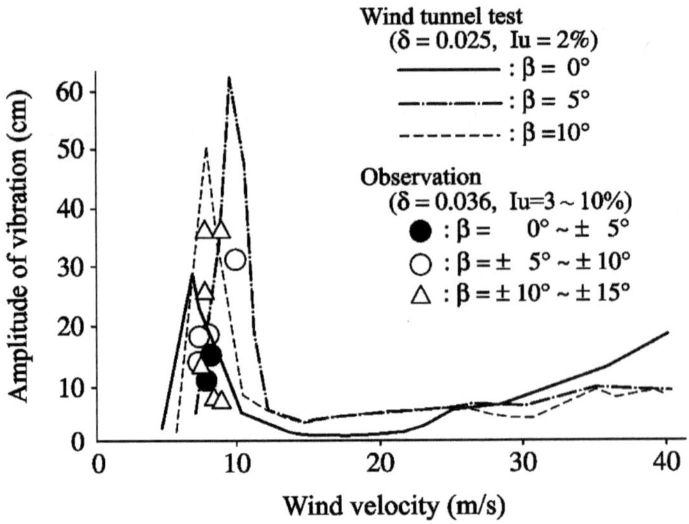

Fig. 9.8 Response from wind tunnel test vs. those from field measurement [6]

(b) It was confirmed that the installed control devices successfully suppressed the wind-induced vibrations to the design level.

9.2.2 Hakucho Bridge

9.2.2.1 Consideration of Vibration Control

The Hakucho Bridge is a suspension bridge with a centre span of 720 m, completed in 1998. It has the portal type towers as shown in Fig. 9.9 and their tower shafts have square-corner-cut cross-section because of the aerodynamic considerations. From the results of the wind tunnel tests, it became evident that the transverse vibration in the first out-of-plane mode would be expected at the wind speed of 6–8 m/s, which could give an adverse effect on the construction work. Active dampers as shown in Fig. 9.10 were installed at the center of the upper middle horizontal strut to reduce the tower top acceleration level down to 0.1 m/s^2 or less, which required the additional damping of 0.18 [8].

9.2.2.2 Vibration Tests

A series of vibration tests was conducted after the tower was completed. The tower was excited in the first bending mode at the frequency of 0.235 Hz using the active

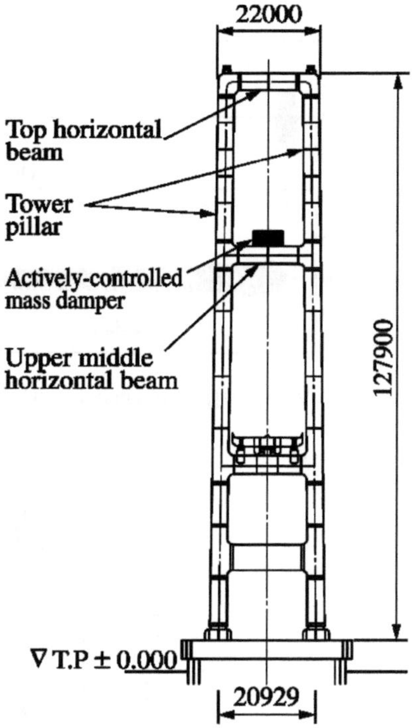

Fig. 9.9 Tower [8]

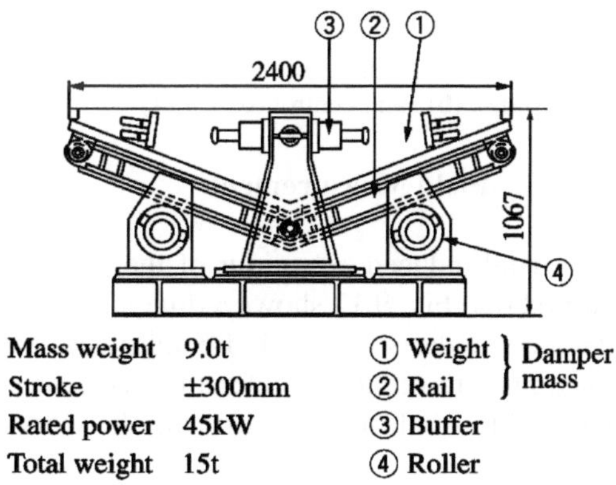

Fig. 9.10 Active control device [8]

damper. The free vibration response, with and without the vibration control, is shown in Fig. 9.11. It was confirmed that the dampers were very effective, providing with the additional damping up to $\delta = 0.21$, as opposed to the target value of $\delta = 0.18$.

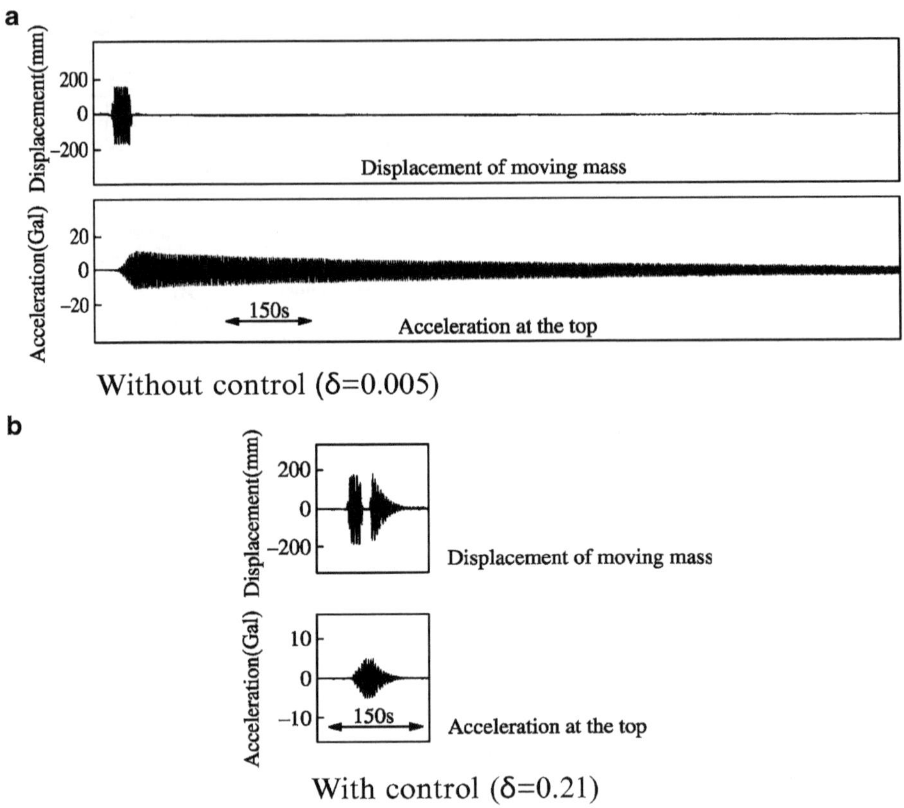

Fig. 9.11 Free vibration response with and without active control devices [8]. (**a**) Without control ($\delta = 0.005$). (**b**) With control ($\delta = 0.21$)

9.2.2.3 Field Measurements

The wind-induced behaviour of the free-standing tower after its completion was carried out. Fig. 9.12 shows a time history of amplitude response with and without dampers under similar wind conditions. When the damper was not working, the maximum tower top acceleration exceeded 0.1 m/s^2, whereas the response was reduced to a half to one-third with the working damper.

9.2.3 Kurushima-Kaikyo Bridges

9.2.3.1 Stability of the Towers After Completion

The Kurushima-Kaikyo Bridges are a group of three successive suspension bridges with the centre spans of 600 m, 1020 m and 1030 m. Their portal towers were

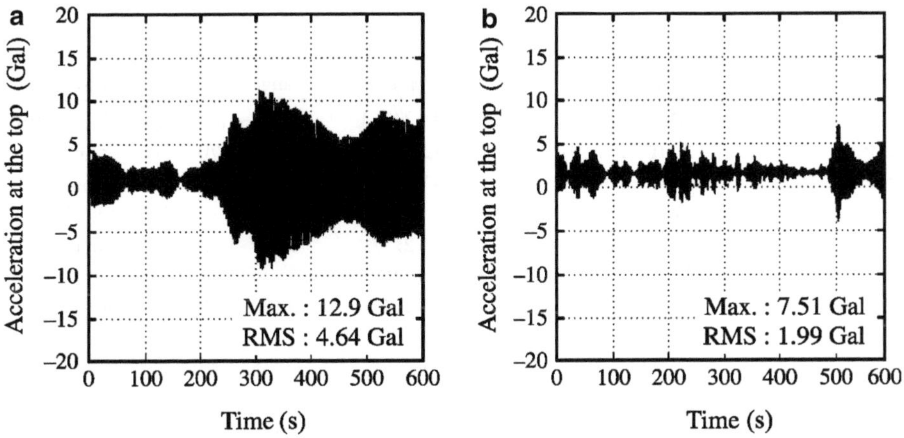

Fig. 9.12 Response under strong wind [8]. (**a**) Without control. (**b**) With control

more flexible than similar towers of existing bridges in Japan, since the stiffening girders of these bridges are streamlined box girders, which are lighter and receive less wind load than conventional truss-stiffened girders. As a result, a possibility of vortex-induced vibration of the tower shafts even after the completion of the bridge was an issue.

The square-corner-cut cross-section was adopted for the tower shafts to reduce the vibration amplitude. The calculated stress in the tower shafts due to vibration, however, needed to be further reduced. For this case, the requirement was satisfied by increasing the cross-sectional area of the tower shaft members.

9.2.3.2 Vibration Control During Construction

It was necessary to suppress the wind-induced vibration of the tower during construction. Vibration was in a number of different modes and also coupled with the movement of the tower crane during the construction period. One problem was that these modes and frequencies changed significantly depending on the construction phase. The active dampers were adopted for this reason. The passive tuned mass dampers were also used particularly for suppressing the higher modes. Table 9.2 presents the dimensions of the vibration control devices used for the towers of the Kurushima-Kaikyo Bridges.

9.2.3.3 Vibration Tests

Vibration tests were conducted using the active damper as an exciter to measure damping when the tower was completed. Table 9.3 shows the damping in terms of logarithmic decrement. In the case of 5P and 9P, the logarithmic decrement in torsional mode changed depending on the amplitude of the vibration.

Table 9.2 Vibration control devices used for the Kurushima-Kaikyo bridges

Tower	Tower height (m)	Vibration control device (mass)	Modes for vibration control[a]	Frequency range	Control theory
2 P	106.45	AMD (9 ton)	O-1, O-2, T-1	0.234~1.67	Quasi-LQ control[b]
3 P	138.95	AMD (20 ton)	O-1, O-2, T-1	0.167~1.704	H^{∞} control
5 P	159.95	AMD (20 ton) TMD (20 ton)	O-1, O-2, I-1, T-1	0.153~1.222	DVFB control[c] H^{∞} control
6 P	137.45	AMD (20 ton) TMD (10 ton)	O-1, O-2, I-1, T-1, etc.	0.197~1.454	Fuzzy active control
8 P	173.45	AMD (22 ton) TMD (20 ton)	O-1, O-2, T-1, etc.	0.133~1.383	Variable gain DVFB control
9 P	170.45	HMD (25 ton) TMD (20 ton)	O-1, O-2, T-1, etc.	0.138~1.458	H^{∞} control

[a]O-1 1st Out-of-plan mode, O-2 2nd Out-of-plan mode, I-1 1st In-plane mode, T-1 1st Torsional mode
[b]Quasi linear quadratic control
[c]Direct velocity feedback control

Table 9.3 Logarithmic decrement of towers in free-standing stage

	Logarithmic decrement		
Tower	1st out-of-plane mode	1st torsional mode	2nd out-of-plane mode
2 P	0.017	0.025	–
3 P	0.010	0.038	0.045
5 P	0.0047	0.017–0.032	
6 P	0.009	0.033	–
8 P	0.008	0.055	0.022
9 P	0.012	0.053–0.018	0.071

The logarithmic decrement of the 1st bending mode was around 0.01 that is the adopted value in design. But it was found greater than 0.01 for other modes. It should be noted that the vibration amplitude was smaller than that assumed in the design and a simple comparison cannot be made.

9.2.3.4 Field Measurements

Figures 9.13 and 9.14 compare the wind-induced vibration observed at the 5P tower top location with the wind tunnel test results, which were obtained when the structural damping was $\delta = 0.01$ and the wind turbulence was $I_u = 10\%$. The observed amplitude was less than predicted by the wind tunnel tests. It was

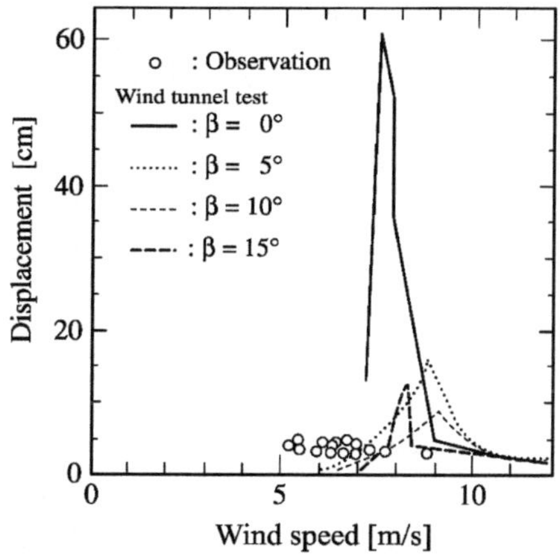

Fig. 9.13 Vortex-induced vibration ($\beta = 0$–$15°$) [9]

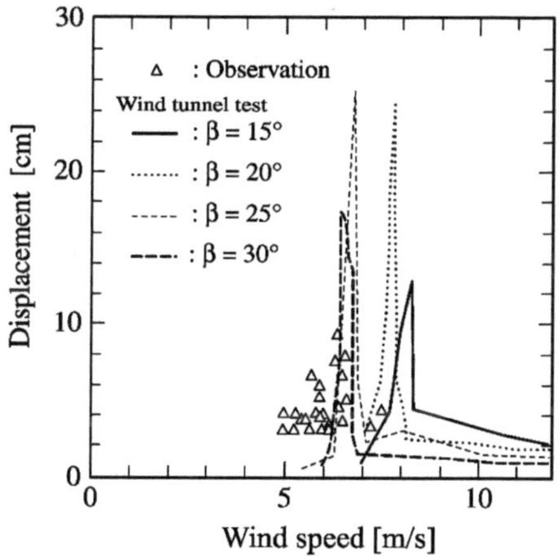

Fig. 9.14 Vortex-induced vibration ($\beta = 15$–$30°$) [9]

probably because the wind in reality was not exactly in the transverse direction against the tower and yet the induced response was found quite sensitive to the change in wind direction.

9.2.4 Other Examples

Tables 9.4 and 9.5 show examples of the mechanical and aerodynamic countermeasures adopted for suppressing the wind-induced tower vibrations [10–16]. Observed frequencies and damping from the field tests are given in parentheses in each table. Also the history of typical bridge tower vibration control is shown as sorted by device types in Table 9.6.

Table 9.4 Mechanical countermeasure

Bridge	Bridge type Center span length (Maximum span length)	Tower type (Height)	Application of counter-measure	Stage[a]	Charateristics of Vibration[b]			
					Vibration mode[c]	Frequency Hz	Effective mass kN/tower	Equivalent mass kN/m/tower
Innoshima	Suspension bridge 3-Continuous-Truss L=770m	Portal 135.85m	1979～1980	Construction	O-1	0.234 (0.238)		118.09
					T-1	0.718 (0.775)		
Oshima (5P)	Suspension bridge Single-box L=560m	Portal 88.35m	1984～1985	Construction	O-1	0.292 (0.304)		
					T-1	1.113 (1.147)		
Ohnaruto	Suspension bridge 3-2 Hinge-Truss L=876m	Portal 125.93m	1980～1981	Construction	O-1	0.34 (0.33)	4958.8	
					T-1	0.74	5654.6	
Kita Bisan-seto (3P)	Suspension bridge 3-Continuous-Truss L=990m	Portal 169.45m	1984	Construction	O-1	0.256	19266.8	
					T-1	0.594	2182	
Shimotsui-seto (2P)	Suspension bridge 3-Continuous-Truss L=940m	Portal 137.63m	1984～1985	Construction	O-1	0.311	7947.8	465.5
					T-1	0.819	10584	1.35E+05
Aratsu	Cable-stayed bridge 3-Continuous Single plane L=185m	Single 60m	1988～	Construction	I-1	0.459		
				Completion	I-1	0.493 (0.42)	2,597	
Bannaguro	Cable-stayed bridge 2-Continuous L=140.4m	Single 45m	1988～	Construction	O-1	1.01	1176	
				Completion				
Yokohama Bay	Cable-stayed bridge 3-Continuous L=469m	Portal 175m	1988～	Construction	O-1	0.274		
					I-1	0.462		
					T-1	0.588		
Rinkai-Tokyo Teleport St. Renraku	Cable-stayed bridge L=146.5m	A-shape 88.2m	1996	Completion	I-1	7.2		
Akashi Kaikyo	Suspension bridge 3-2 Hinge-Truss L=1,991m	Portal 297m	1993～	Construction	O-1	0.127 (0.126)		
					O-2	0.677 (0.673)		
					T-1	0.473 (0.471)		
				Completion	O-1	0.442		745.78
					T-1	0.747		813.4
Tsurumi Tsubasa	Cable-stayed bridge 3-Continuous L=510m	A-shape 180m	1992～ 1993	Construction	O-1	(0.280)		

a Construction: Free standing
b ( ): field test result
c Vibration mode: O-1: 1st out-of-plan mode, O-2: 2nd out-of-plan mode, I-1: 1st in-plane mode, T-1: 1st torsional mode

(continued)

	Vortex induced oscilation			Galloping		Countermeasure		
Logarithmic decrement ratio	Bending Mode: Bridge axis	Bending Mode: Transvers axis	Torsional mode	Bridge axis	Transvers axis	Device	Mass kN	Effect of countermeasure
(0.008~0.015)	○					S B	22.050	Maximum amplitude of tower top 77cm→40cm
			○				54.684	
(0.016)	○					S B	23.912	Logarithmic decrement ratio δ=0.016→0.042
(0.032)			○				127.4	Logarithmic decrement ratio δ=0.032→0.064
0.01	○					O D	196 /shaft	Increse of logarithmic decrement ratio δ=0.1 (at the ampletude of 20cm)
0.01			○					
0.01	○					O D	490 /shaft	Maximum amplitude of tower top 1.7m→0.2m
0.01			○					
0.01	○					O D	490 /shaft	Maximum amplitude of tower top 1.74m→0.37m
0.01			○					Maximum amplitude of tower top 0.26m→0.17m
	○							
0.01 (0.009)		○			○	T M D	35.084	Logarithmic decrement ratio δ=0.009→ δ=0.121
0.01 (0.013~0.023)		○				T M D	17.64	Logarithmic decrement ratio δ=0.023→ δ=0.285
		○				T M D		
0.01	○					T M D		
0.01		○						
0.01			○					
0.01		○			○	T M D		
(0.0067)	○					T M D	2146.2	Logarithmic decrement ratio O-1: δ=0.0067→0.036 O-2: δ=0.038→0.080 T-1: δ=0.028→0.075
(0.038)	○							
(0.028)		○				H M D	548.8	
0.02	○					T M D	1940.4	
0.02		○						
(0.012)	○					H M D	98	Logarithmic decrement ratio O1: δ=0.012→0.475

S B : Sliding block
O D : Oil-damper
T M D : Tuned mass damper
H M D : Hybrid mass damper
A M D : Active mass damper
T S D : Tuned sloshing damper
T L D : Tuned liquid damper
I M D : Impact mass damper

Bridge	Bridge type Center span length (Maximum span length)	Tower type (Height)	Application of counter-measure	Stage 1)	Charateristics of Vibration 2)			
					Vibration mode 3)	Frequency Hz	Effective mass kN/tower	Equivalent mass kN/m/tower
Rainbow (Daiba-side tower)	Suspension bridge 3-2 Hinge-Truss L=570m	Portal 117.4m	1991	Construction	O-1	0.278	13602.4	227.4
					T-1	1.025	19208.0	
Hakucho Ohashi (Iwaizu-side tower)	Suspension bridge 3-2 Hinge-Box L=720m	Portal 131m	1992	Construction	O-1	0.227 (0.228)		
					O-2	1.219 (1.247)		
					T-1	1.139 (1.195)		
Hakucho Ohashi (Jinya-side tower)	Suspension bridge 3-2 Hinge-Box L=720m	Portal 127.9m	1992～1993	Construction	O-1	0.229 (0.235)	5468.4	64.3
					O-2	1.263 (1.290)		
					T-1	1.178 (1.220)	7389.2	62.4
Meiko Chuo (Central)	Cable-stayed bridge 3-Continuous L=590m	A-shape 195m	1994～	Construction	O-1	0.159		
					I-1	0.443		
					T-1	0.943		
Nakajima	Cable-stayed bridge 3-Continuous L=163m	Portal 71m	1996	Construction	O-1	0.494		26.6
					O-2	2.999		25.9
1st Kurushima Kaikyo (2P)	Suspension bridge 3-2 Hinge-Box L=600m	Portal 112.25m	1996～1997	Construction	O-1	0.264	6399.4	
					T-1	0.959	6066.2	
1st Kurushima Kaikyo (3P)	Suspension bridge 3-2 Hinge-Box L=600m	Portal 144.75m	1995	Construction	O-1	0.227	5331.2	157.8
					I-1	0.670	7810.6	155.8
					T-1	0.803	7134.4	139.2
Ikuchi	Cable-stayed bridge 3-Continuous L=490m	A-shape 120.3m	1989	Construction	O-1	0.255 (0.249)	4909.8	95.6
Sakitama (downstream-side)	Cable-stayed bridge 2-Continuous L=190m	Single 80m	1990	Construction	O-1	0.490	2126.6	90.5
					I-1	0.529	1489.6	74.3
Sakitama (upstream-side)	Cable-stayed bridge 2-Continuous L=190m	Single 80m	1990	Construction	O-1	0.625	4410.0	
					I-1	0.439	2058.0	
Sanuki Huchuko	Cable-stayed bridge 2-Continuous L=130m	Single 53m	1991～	Construction	I-1	0.574	659.5	
				Completion	I-1			
Akinada (3P)	Suspension bridge 3-2 Hinge-Box L=742m	Portal 119.45m	1996～	Construction	O-1	0.222 (0.221)		
					I-1	1.284 (1.303)		
					O-2	1.284 (1.303)		
					T-1	1.145 (1.154)		

(continued)

	Vortex induced oscilation			Galloping		Countermeasure		
Logarithmic decrement ratio	Bending Mode		Torsional mode	Bridge axis	Transvers axis	Device	Mass	Effect of countermeasure
	Bridge axis	Transvers axis					kN	
(0.012)	○					HMD	19.6	Maximum amplitude O-1: →1/4
			○			TMD	98	
(0.005)	○					HMD	78.4	Logarithmic decrement ratio → δ =0.18
(0.029)								
(0.018)			○					
(0.005)	○					HMD	88.2	Maximum amplitude →under 10gal
(0.052)								
(0.013)			○					
0.01	○					HMD	137.2	
0.01								
0.01								
(0.031)	○					HMD	34.3	Logarithmic decrement ratio δ =0.031→0.119
	○							
(0.013)	○					HMD	88.2	Logarithmic decrement ratio O-1: δ =0.013→ δ =0.073 I-1: δ =0.092→ δ =0.126
(0.092)			○					
(0.01)	○					AMD	196	Logarithmic decrement ratio O-1: δ =0.01→ δ =0.033
0.01								
0.01			○					
0.0095 (0.013)	○					TSD	0.94% of Effective Mass	Logarithmic decrement ratio δ =0.013→ δ =0.21∼0.23
(0.031)	○					TSD	Out-of-plane:12.5 In-plane: 14.1	Logarithmic decrement ratio O-1: δ =0.031→ δ =0.097 I-1: δ =0.035→ δ =0.068
(0.035)		○						
(0.03∼ 0.06)	○					TLD	14.7	Increse of Logarithmic decrement ratio △ δ =0.06∼0.13
(0.02∼ 0.05)		○						
(0.022)	○					IMD	9.0	Logarithmic decrement ratio δ =0.022→ δ =0.226
	○							
(0.012)	○					TMD	107.8	Logarithmic decrement ratio δ =0.012→ δ =0.147
(0.0098)				○			39.2	Logarithmic decrement ratio δ =0.0098→ δ =0.216
(0.044)	○					TMD	156.8	Logarithmic decrement ratio δ =0.044→ δ =0.176
(0.036)			○					Logarithmic decrement ratio δ =0.036→ δ =0.399

Table 9.5 Aerodynamic countermeasure

Bridge	Bridge type Center span length (Maximum span length)	Tower type (Height)	Application of counter-measure	Stage[a]	Charateristics of Vibration[b]			
					Vibration mode[c]	Frequency Hz	Effective mass kN/tower	Equivalent mass kN/m/tower
Katsushika Harp	Cable-stayed bridge 4-Continuous Single plane L=220m	Single 68.2m	1985∼	Construction				
				Completion	I-1	0.568		
Higashi-Kobe	Cable-stayed bridge 3-Continuous L=485m	Portal 145.3m	1988∼ 1990	Completion	I-1	0.25		
				Completion	I-2	1.05		
Nakagawa	Cable-stayed bridge L=285m	A-shape 106.348m		Construction	O-1	0.381		13.4
Akashi Kaikyo	Suspension bridge 3-2 Hinge-Truss L=1,991m	Portal 297m	1998∼	Completion	O-1	0.442		
				Completion	T-1	0.747		
Tatara	Cable-stayed bridge 3-Continuous L=890m	Inverted Y-shape 220m	1999∼	Completion	O-1	1.109		400.8
Uragami	Cable-stayed bridge L=51m	Single 24m		Completion	I-1	1.14		4.9
Nakajima	Cable-stayed bridge 3-Continuous L=163m	Portal 71m	1994∼	Completion	I-1	0.88		57.7
Megami	Cable-stayed bridge 3-Continuous L=480m	Portal 170m	1996∼	Completion	I-1	0.481		
Sugawara-Shirokita	Cable-stayed bridge 3-Continuous L=238m	Single 43m	1989∼	Completion	I-1	1.06		86.1
Hamada Marine	Cable-stayed bridge 2-Continuous L=200m	A-shape 89m	1999∼	Construction	O-1	0.368 (0.374)		
					I-1	0.942 (0.967)		

a Construction: Free standing
b ( ): field test result
c Vibration mode: O-1: 1st out-of-plan mode, O-2: 2nd out-of-plan mode, I-1: 1st in-plane mode, T-1: 1st torsional mode

(continued)

	Vortex induced oscilation			Galloping		
Logarithmic decrement ratio	Bending Mode		Torsional mode	Bridge axis	Transvers axis	Countermeasure
	Bridge axis	Transvers axis				
	○	○				Arc-like deflecter + side plate
0.01		○			○	Arc-like deflecter
0.01					○	Cover plate
0.01						
0.01				○		Cover plate
0.02	○					Corner-cut
0.02			○			
0.02				○		Corner-cut
0.01					○	Faring plates
0.02		○				Faring plates
0.02		○			○	Faring plates
0.01		○			○	Slit
(0.006～0.007)	○			○		Triangular recess
(0.02～0.04)		○			○	

Table 9.6 History of vibration control used for bridge towers in Japan

	Countermeasure	1960		65			70		75			80
Mechanical countermeasure	Sliding block	Force					Kanmon				Innoshima	
	Oil-damper+Block										Innoshima	
											Ohnaruto	
	Tuned mass damper											
	Tuned sloshing damper／Tuned liquid damper											
	Impact mass damper											
	Hybrid mass damper											
	Active mass damper											
Aerodynamic countermeasure	Deflector											
	Rectangular recess											
	Triangular recess											
	Slit											
	Plates											
	Horizontal strut											

▬ : Construction stage ⟶ : Completion stage

(continued)

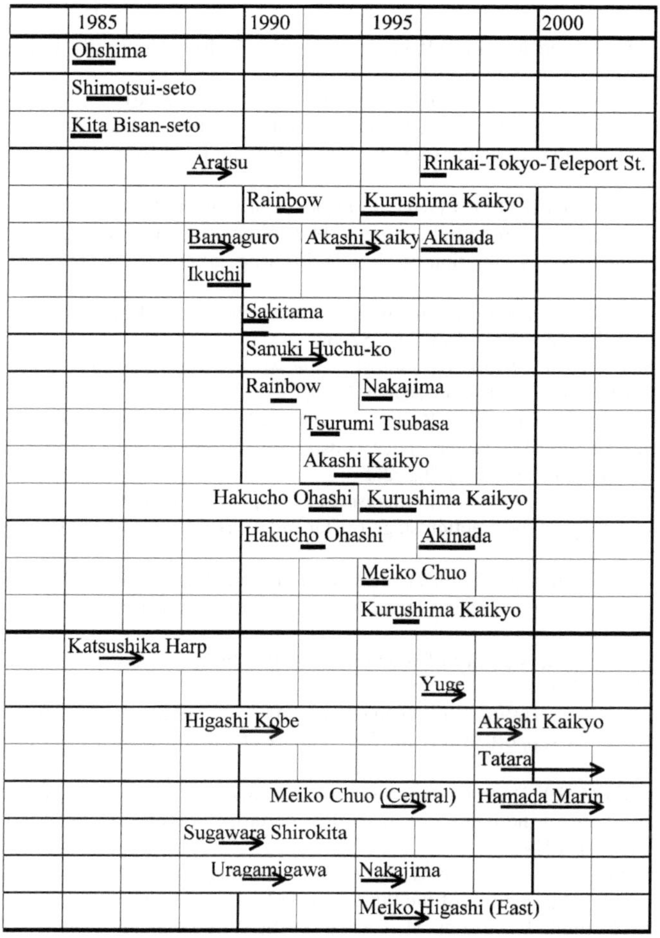
1985
1990
1995
2000
Ohshima
Shimotsui-seto
Kita Bisan-seto
Aratsu
Rinkai-Tokyo-Teleport St.
Rainbow
Kurushima Kaikyo
Bannaguro
Akashi Kaiky
Akinada
Ikuchi
Sakitama
Sanuki Fuchu-ko
Rainbow
Nakajima
Tsurumi Tsubasa
Akashi Kaikyo
Hakucho Ohashi
Kurushima Kaikyo
Hakucho Ohashi
Akinada
Meiko Chuo
Kurushima Kaikyo
Katsushika Harp
Yuge
Higashi Kobe
Akashi Kaikyo
Tatara
Meiko Chuo (Central)
Hamada Marin
Sugawara Shirokita
Uragamigawa
Nakajima
Meiko Higashi (East)

References

1. Scruton C (1981) An introduction to wind effect on structures. Oxford University Press, Oxford
2. Bridge and Wind Editorial Group (1994) Bridge and Wind, Kyoto (in Japanese)
3. Okajima A, Ueno H, Abe A (1991) Influence of Reynolds number on flow and aerodynamic characteristics of bluff bodies with rectangular section of cut corners (in Japanese). J Wind Eng JAWE (49):1–13
4. Ohe S, Akiyama H, Mori K, Hata K (1997) Dynamic response of towers of Akashi Kaikyo Bridge. In: Proceedings of the ASCE structural congress 1997, Wind effect on bridge, pp 1–5
5. Fujino Y (1990) Countermeasures of vibration of civil engineering structures (in Japanese). J Wind Eng JAWE (44):53–69
6. Hata K, Tatsumi M (1998) Vibration control of the main towers of the Akashi Kaikyo Bridge. In: Proceedings of the IABSE symposium, Kobe, pp 57–62
7. Tatsumi M, Mori K, Hata K (1993) Vibration control of the Akashi-Kaikyo Bridge's main towers (in Japanese). Honshi Technical Report BOEA 17(68):16–27
8. Takahashi M, Nishimoto S, Hashimoto K, Koike Y, Mutaguchi M, Ueshima H (1994) Vibration control of Hakucho Bridge tower under construction using actively-controlled mass damper. J Robot Mechatron JSME 6(3):262–265
9. Suzuki S, Sasaki M, Yamaguchi K (1996) Countermeasures to suppress vibration of the towers in the Kurushima Kaikyo Bridge in the erection stage (in Japanese). Honshi Technical Report BOEA 20(80):27–39
10. Yoshimura T, Inoue A, Kaji K, Savage M (1989) A study on the aerodynamic stability of the Aratsu bridge. In: Proceedings of the Canada-Japan workshop on bridge aerodynamics, pp 41–50
11. Yoshimura T (1992) Aerodynamic stability of four medium span bridge in kyushu district. J Wind Eng Ind Aerod 42:1203–1214
12. Saito T, Ito M, Yamaguchi H, Eya S (1984) Yokohama Bay Bridge – Use of a dynamic damper for aerodynamic stability of its support tower. In: Proceedings of the 4th US-Japan bridge workshop, San Diego
13. Miyata T, Sato H, Kitagawa M (1993) Design considerations for superstructures of the Akashi Kaikyo Bridge. In: Proceedings of the international seminar on utilization of large boundary layer wind tunnel, Tsukuba, pp 121–139
14. Honda A, Tatsumi M, Hata K, Ohnichi E (1995) Aerodynamic characteristics of the tower of the Akashi-Kaikyo bridge. In: Proceedings of the 9th ICWE, New Delhi, pp 803–814
15. Kamei M, Kawakami M, Ikushima K, Hosomi M, Tanaka H (1993) Aerodynamic stability of Nakajima bridge. In: Proceedings of the EASEC-4
16. Saito T, Shiraishi N, Fujisawa M (1987) On aerodynamic improvement of pylon of cable-stayed bridge with single cable plane. In: Proceedings of the international seminar on wind engineering, Innsbruck

Chapter 10
Cable Vibrations and Control Methods

10.1 Introduction

The development of cable-stayed bridge as a structural choice for medium to long span bridges has been remarkable through the closing decades of the last century. As an essential component of the bridge structure, stay cables play an important role in the dynamic behaviour of cable-stayed bridges. Cables are extremely vulnerable to wind excitation mainly due to its low mechanical damping. Many efforts have been made during the past years to clarify the mechanisms of, and find solutions to, various types of wind-induced cable vibrations to alleviate engineering problems. Furthermore, with a rapid development of span length in cable-stayed bridges, even new types of instability particularly of the inclined cables have been identified, such as the rain–wind vibration, high-speed vortex excitation, and the dry inclined cable galloping, which have been all new challenges to bridge engineers. The purpose of this chapter is to attempt a comprehensive state-of-the-art review of various types of wind-induced cable vibrations.

The wind-induced cable vibrations can be categorized into several groups, depending mainly upon their excitation mechanisms but also somewhat upon their historical context. They can be as follows:

(a) Vortex-induced vibration, or aeolian oscillation
(b) Buffeting due to wind gust
(c) Classical galloping typically observed in iced cables
(d) Wake interference, or wake galloping and resonant buffeting
(e) Parametric excitation
(f) Reynolds number related drag instability
(g) Rain–wind induced vibration
(h) High-speed vortex excitation
(i) Dry inclined cable galloping

The first two types of vibration are generally small in amplitude and possible fatigue failure would be the only engineering concern associated with them.

Y. Fujino et al., *Wind Resistant Design of Bridges in Japan*,
DOI 10.1007/978-4-431-54046-5_10, © Springer 2012

The last three types are mainly related to inclined cables, such as bridge stay cables, and the rain–wind induced vibration is the one most frequently observed in reality. The excitation mechanism of rain–wind vibration has been made clearer in recent years, and some effective control methods have been successfully applied in practice. However, the last two types of motion are still less understood and would require more intensive research attention. The dry inclined cable galloping, in particular, would cause a very strong concern since it is suggested that they can result in undesirably large amplitude motion.

In this chapter, a brief review of various types of wind-induced cable vibration together with its control methods is attempted.

10.2 Various Types of Wind-Induced Cable Vibrations

10.2.1 Vortex-Induced Vibration, or Aeolian Oscillation

This is a small amplitude vibration caused by vortex shedding and takes its name from the æolian harp, an ancient Greek instrument functioned on the same mechanical principle. The basic mechanism of excitation is considered f_V to be a resonance to the frequency of vortex shedding whose frequency is given by

$$f_V = St \cdot \frac{\bar{U}}{d} \tag{10.1}$$

where $\bar{U} =$ the mean wind speed, $d =$ the outer diameter of the cable. St is the Strouhal number, which is approximately 0.19–0.20 for a circular cross-section and the Reynolds number of -10^5 or less. As a typical example, if $d =$ 0.16 m and $St =$ 0.19 are assumed, for the mean wind speed of $\bar{U} = 5$–25 m/s, corresponding vortex shedding frequency is in the range of $f_V =$ 10–50 Hz. Since the cable's fundamental frequency is often in the range of 0.2–2 Hz, the resonance occurs only with much higher harmonic modes with vibration frequencies of 10–50 Hz, where the mechanical self-damping is likely to be quite high. The maximum vibration amplitude usually does not exceed one cable diameter, peak-to-peak. Wind turbulence generally tends to reduce the response amplitude, even down to a half compared to the exposure to smooth air flow [1].

The same mechanism of wind excitation can cause much more serious engineering problems for towers, chimneys and bridge road decks. Both experimental and analytical investigation has been devoted to this topic and their outcomes have been reflected to the design codes in practice. However, vortex shedding excitation of cables is not a major concern for engineers except that it may cause fatigue damages near the cable clamps. For the case of power transmission lines, it is a common practice to install the Stockbridge-type dampers [2] or helical surface rods around the conductor. It is noteworthy that the Stockbridge dampers also contribute in reducing the twist of cables but they also have serious problems of failure due to fatigue. For bridge cables, the viscous dampers are more commonly used. A novel device, the passive damper cable, has been proposed recently [3]. The slack nature of the damper cable exhibits large static hysteresis caused by inter-strand friction of the cable during bending motion, which is said to dissipate energy effectively.

10.2.2 Buffeting due to Wind Gust

Gust induced random vibration is generally not a very serious concern for structural cables except for the power transmission lines. There are a few fundamental references, such as a design guideline prepared by the ASCE Committee [4] in this respect. The basic idea of its recommendation is to come up with a gust response factor for the prediction of dynamic response by following a conventional buffeting analysis in the frequency domain. The procedure is based upon an earlier publication by Davenport [5]. Further improvement of its contents should include particularly the span-wise force correlation, wind yaw angle effects, and the longitudinal loads along the cable span.

For bridge cables, buffeting is generally a less serious problem. High tension of bridge stay cables generally helps to limit the amplitude of buffeting. Sometimes buffeting is observed as a result of wake interference, where the existence of upstream objects is the cause of disturbance causing instability of downstream cables. Electric wires along the main cables of the Golden Gate Bridge exhibited vibration a couple of decades ago, which was probably buffeting of this kind. It has been pointed out [6] that wake buffeting could produce aerodynamic instability in bridges when there are two parallel cable planes. If the time required for the wind flow to travel from the upstream plane to the downstream plane of cables is equal to a half of a cycle of the torsional vibration of the deck, it is conceivable that cable buffeting can enhance the deck vibration and cause the bridge instability, though, in the authors' knowledge, this has never been an issue in reality.

10.2.3 Galloping of Iced Cables

The accretion of ice on a conductor modifies its cross-sectional geometry and hence its aerodynamic characteristics. This may result in an aerodynamically induced instability called galloping. The motion is principally in vertical direction of a horizontal cable, with a low frequency of typically less than 1 Hz, and a large amplitude such as 10–20 m, large enough to cause serious operational problems. The primary reason of this mechanism is a significant negative slope in the lift force curve against the angle of attack, α, which gives the exciting lift force in the same direction as the cable motion. Thus, energy from the surrounding airflow is fed continuously to the system, leading to an unstable motion, which is somewhat similar to flutter. The instability criterion given by Den Hartog [7] is

$$\frac{\partial C_L}{\partial \alpha} + C_D < 0 \tag{10.2}$$

where C_L and C_D are the lift and drag force coefficients, respectively.

However, strictly speaking, since the drag force also slightly depends upon the angle of attack, the path of the cable motion tends to follow an elliptical trace. Furthermore, the instability is known to be triggered by aerodynamic coupling with torsion [8]. On the other hand, the static deflection due to drag force and pitching moment under high wind speed can effectively change the aerodynamic characteristics of the cross-section, which could even stabilize the cable [9]. A similar behaviour for the case of non-circular cable sections has been also discussed by Ito et al. [10]. Since the dynamic amplitude of galloping tends to be so large, the cable behaviour becomes strongly nonlinear. Unlike torsional flutter, wind turbulence does not necessarily work as a stabilizing factor for galloping [9]. Engineering problems are not restricted to the power conductors. Some studies [11] have shown that the same mechanism of excitation may have resulted galloping of heavy stay cables, leading to the collapse of a guyed tower. It has been reported that the violent vibration of bridge stay cables observed at the Ørsund link between Denmark and Sweden in February 2004 was likely to be galloping caused by snow and sleet accumulation on cable surface [12].

A comprehensive list of the references related to the research on galloping has been presented by Lilien [8]. The analysis itself of galloping does not involve too much theoretical difficulties in principle. However, it needs to be supported by reliable aerodynamic data for a range of realistic natural ice shapes formed under various weather conditions of interest. The collection of this information will greatly assist the theoretical study of galloping.

10.2.4 Vibrations Caused by Wake Interference

The dynamic behaviour of structures excited by wind can be drastically altered by their proximity to its neighbouring structures. These mechanisms that lead to aerodynamic excitation do not of course exist for a single isolated cable. Much research on this category of fluid-structure interaction has been motivated by problems encountered with the closely spaced tubes exposed to internal flow of heat exchangers and also with bundled conductors used in high voltage electric power transmission lines. There are, however, some other examples of proximity effects also related to spaced cables, slender towers and chimneys. Due to the complexity of flow pattern around closely spaced structures, several different driving mechanisms can arise. Zdravkovich [13], for example, has extensively

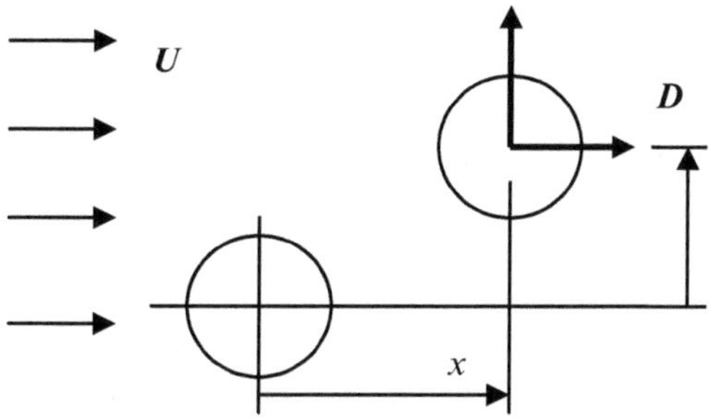

Fig. 10.1 Two cylinders staggered

studied various cases of aerodynamic interaction between two circular cylinders (Fig. 10.1) located close to each other.

Of particular importance, in the present context, is the case when two circular cylinders are placed in staggered arrangements. The predominant response is (1) resonant buffeting, where the vortex wake of an upstream body is resonant with the natural frequency of a structure submerged in the wake; and (2) wake galloping, where lift and drag forces in the wake shear flow lead to a coupled instability of a submerged cylinder.

Wake interference galloping occurs when two cylinders are either closely or largely spaced. For close spacing, the flow around two cylinders is significantly altered by aerodynamic interference between two cylinders. Stay cable vibrations of the Yobuko Bridge [14] was of this kind. Extensive studies on instability of closely spaced cylinders have been conducted, particularly for heat exchanger bundles [15] and bridge stay cables. Instability is observed in the range of $-2<y/d<2$ and $1<x/d<4$. It starts at the critical reduced velocity of $(\bar{U}/fd)_{cr} \approx 40$, which typically corresponds to the flow speed of 5–20 m/s, and follows generally an elliptical trajectory with the maximum amplitude of less than $3d$. The motion is known to be sensitive to the Scruton number, $Sc = m\varsigma/(\rho d^2)$ and becomes hardly recognizable at $Sc>50$. To reduce the vibration amplitude, the installation of viscous dampers at cable ends or connecting the adjacent cables by cross-ties have been practiced. However, these methods are not effective for longer cables.

As spacing increases, the interference effects diminish until the next "large spacing" instability range, $8<x/d<20$, is reached, where the interference effects are only on the downstream cylinder and the flow around the upstream structure is no longer affected by the second cable behaviour. The interference effects of largely spaced structures have attracted less research attention except for the wake galloping of bundled conductors and a good comprehensive reference of both analyses and test results is presented by Wardlaw [16]. The transmission of electric power in very high voltage requires suspension of conductors in bundles to avoid a corona discharge to ground. The separation of parallel conductors, typically in the range of 10–20 conductor diameters, is maintained by the use of spacers that usually divide the span into 50–60 m sub-spans. This becomes the reason of having a distinctively different type of cable vibration called sub-span oscillation caused by wake interference.

If the upstream cable is fixed, the downstream conductor would move in an elliptic path with the long axis nearly horizontal. Amplitudes become large enough that conductors could clash. Although it is the downstream conductor that becomes aerodynamically unstable, the field observations report the existence of anti-phase motion of a conductor pair with both cables having similar amplitudes and with the frequency in the range of 1–4 Hz. This is in contrast to the vortex shedding excitation which takes place at higher frequencies of 10–50 Hz, and the large amplitude galloping of iced conductors at the frequency less than 1 Hz.

The aerodynamic mechanism of this vibration was extensively investigated in the 1970s and is fairly well understood by now. Research activities included aerodynamic force measurements [17] and mathematical analyses (for example [18]). It has been found that the presence of wind velocity gradient across the wake of the windward object would induce position-dependent lift and drag forces on the downstream cable. The most sensitive region for this excitation is approximately at a quarter of the width of the wake from its outer edge shear layer, where the lift force reaches its maximum.

A conventional practice of suppressing this type of motion is with spacer-dampers that are flexible and often installed at unequal intervals. However, the impact of the damper provided by the spacers on the instability is still not clear. Opposite conclusions were drawn from different studies. The use of low level damping has been recommended by a Hydro-Quebec study [19], while Price and Paidoussis [20] advised to use very high level of damping.

The unexpected wake-induced flutter was observed on the hanger ropes of the Akashi Kaikyo Bridge in Japan [21]. Cable distance was $9d$ for this case, normal to the bridge axis. Spiral ropes of 10 mm in diameter were used to wind up the hanger ropes to suppress both the wake-induced flutter and vortex shedding excitation effectively.

10.2.5 Parametric Excitation

Vibrations of stay cables can be excited by motion of the anchorage points, as it was first pointed out by Kovács and Leonhardt [22]. Detailed analytical as well as experimental study on the dynamic interaction of cable-girder system was conducted by Fujino et al. [23, 24]. This is sometimes observed in cable-supported structures such as cable-stayed bridges and guyed towers. One of the important points in particular is the fact that a cable can be excited not only with its natural frequency but also with the excitation frequency which is two times its own natural frequency.

The most likely cases are when the excitation frequency is approximately twice or equal to the first natural frequency of the cable. This is only an indirect aerodynamic excitation of cables. Nevertheless, it can be a serious issue and should be briefly mentioned here. Because of its low structural damping, the cable motion can develop to large amplitude even if the end excitation is small. Also, since the bridge stay cables present a wide range of natural frequencies, there is a good possibility of some of the cables resonating with the movements of the deck or pylons due to wind or traffic loads. It should be noted also that the cable vibration may not be only in its fundamental mode. Parametric excitation of bridge stay-cables including nonlinear characteristics and varieties of parameter changes is an interesting research topic but is not fully investigated yet [25]. However, it is also said that the stay cable vibration is likely to stay linear because of very small sag and small end motions [26]. It has been reported that the stay cables of the Second Severn Crossing experienced severe vibration apparently because of this reason. When the cross-ties were installed to alleviate the cable vibration, the bridge deck started vibrating, since now it has lost an effective TMD [27]. For the case of the Normandy Bridge, the problem was removed by adding cross-ties between stay cables to avoid resonant vibrations [6].

10.2.6 Reynolds Number Related Drag Instability

This phenomenon was observed several decades ago on a stranded wire conductor crossing the Severn River in England. The vibration was severe enough to cause numerous electrical faults due to conductor clashing. There was apparently no vertical conductor motion associated with it. A very interesting fact was that the wind speed and direction at which the instability occurred were confined to rather narrow bands. The physical dimensions involved in the case are as follows: main span length = 1,619 m, the sag = 80.5 m, cable diameter = 43 mm and cable mass = 6.40 kg/m. Vibration frequency was found in the range of 0.128–0.130 Hz, which approximately corresponds to the first asymmetric mode in lateral sway. The vibration was observed at the mean wind speed 13–15 m/s and most of the time the mean wind direction was 10–25° deviated from the normal to the span. Results from extensive studies [28, 29] pointed out that this unusual instability was due to the sensitivity of the drag force to the Reynolds number.

As it has been well-recognized, the aerodynamics of cylindrical bodies is highly sensitive to the change of flow speed near the critical Reynolds number. Though the critical Reynolds number is influenced by the roughness and texture of the body surface as well as the flow turbulence, it is characterized by a sudden drop of the drag force with the increase in wind speed. If the cable is swinging back and forth parallel to the wind direction resulting in the change of the relative wind speed at this sensitive speed range, it is possible to have the induced aerodynamic drag force acting in the same direction as the body motion, and thus generate the negative aerodynamic damping. In case of a smooth circular section, this "drag crisis" occurs at the Reynolds number range of $2\text{–}5 \times 10^5$. Considering the cable diameter to be the order of 50–150 mm, possibility of having this instability is at the wind speed of 20–60 m/s, which is possible to take place but has never been reported as a problem in reality.

For the case of the Severn crossing, the helical stranding of conductors featured two influences on the aerodynamic characteristics. First, the critical Reynolds number for the conductor was about an order of magnitude lower than that of the smooth circular cylinder, down to $2.6–4.0 \times 10^4$ because of the characteristic roughness. Second, in oblique winds, the lay of the strands streamlined the flow on surface in which the wind was more parallel to the stranding and roughened the opposite surface. This deflected the flow streamlines and generated a mechanism for lift. The effect was further amplified at the critical Reynolds number when the mechanism of drag damping was also active. This mechanism caused a large amplitude motion and the clashing. As a countermeasure, the conductor was wrapped in PVC tape to eliminate the undesirable impact of the stranded surface.

Considering the range of Reynolds number and possible cable dimensions, similar instability is conceivable for bridge stay cables, too. However, so far no particular case other than this has been attributed to the same mechanism.

10.2.7 Rain–Wind Induced Vibration

Rain–wind induced cable vibration has been most frequently reported for bridge cable vibration. It was first addressed by Hikami [30] when the stay-cables of the Meiko-Nishi Bridge in Japan experienced annoying vibrations. The most curious part was that it was observed under certain wind conditions, but only when it was raining. The observed vibration amplitude was up to approximately $2d$, d being the cable diameter, which was typically 14 cm, and under the wind of 8–14 m/s. A simple analysis led to the conclusion that the observed vibration was not any of the known types. The observed frequency was 1–3 Hz, which was well below the Strouhal frequency for the vortex shedding excitation. The cables were too far apart in distance to cause any aerodynamic interference. The observed cable vibrations were hence assumed to be a new type of instability, which is caused by the combined action of rain and wind.

After this new type of excitation mechanism was reported, it became clear that, in fact, there had been some other cable vibrations reported earlier for other bridges, which could have been classified into the same category. Some of the reported cases of this type of cable vibration are given in Table 10.1 together with other field observations.

The cause of this vibration is reported to be in two steps. First, the formation of upper water rivulet on the cable surface seems to be a key factor [30, 31] (Fig. 10.2). It is formed as a result of a sensitive equilibrium between gravity, capillary and aerodynamic forces. The water rivulet effectively alters the geometrical cross-section of the cable and hence the aerodynamic forces on it. Depending on the location and size of the rivulet, it tends to give a negative slope of lift curve against the small change in the angle of attack and also significantly reduce the drag force,

Table 10.1 Observed rain–wind vibrations

Bridge	Country	Year	References
Köhlbrand	Germany	1974	Ruscheweyh and Hirsch [70]
Brottonne	France	1977	Wianecki [71]
Meiko-Nishi	Japan	1984	Hikami [30], Hikami and Shiraishi [51]
Farø	Denmark	1985	Langsø and Larsen [50]
Tempozan	Japan	1986	Miyasaka et al. [72], Oshima and Nanjo [73]
Aratsu	Japan	1988	Yoshimura et al. [48, 74]
Ben Ahin	Belgium	1988	Lilien et al. [75], Cremer et al. [76]
Burlington	Vermont, USA	1990s	Virlogeux [6]
Glebe Island	Australia	1990s	Virlogeux [6]
Nampu	China	1992	Cheng [77]
Yangpu	China	1995	Gu et al. [78]
Erasmus	Holland	1996	Geurts et al. [79]
Ørsund	Denmark/Sweden	2001	Larsen and Lafrenière [12]
Cochrane	Alabama, USA	2002–2004	Irwin [80], Irwin et al. [58]

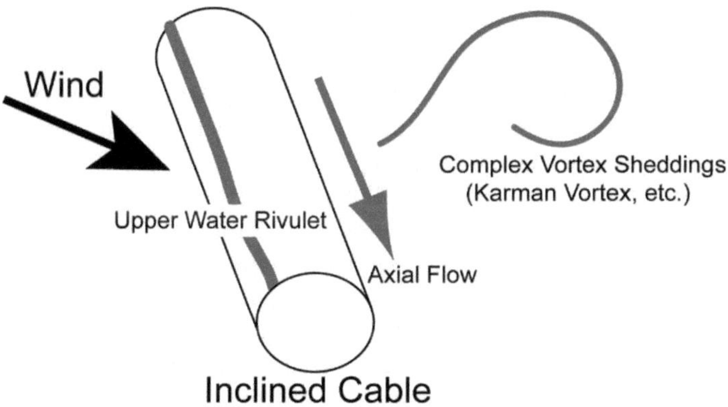

Fig. 10.2 Factors considered in aerodynamics of an inclined cable

which results in the Den Hartog type galloping instability [32]. Also, once the cable is set to motion, the upper rivulet tends to oscillate along the cable surface in circumferential direction and this motion can be aerodynamically coupled with the flexural oscillation of the cable, making the modal aerodynamic damping negative. Naturally, it is expected to intensify the vibration [30, 32, 33]. Contrary to this explanation, Bosdogianni and Olivari do not believe the motion of liquid rivulets has any influence on the instability [34]. More research on the second triggering factor, in fact, identified three fundamentally different excitation mechanisms associated with along-wind, cross-wind and mainly across-wind cable vibration under rain and wind conditions [35, 36]. They essentially depend on the cable orientation and wind speed. An approximate method to estimate the amplitude of rain–wind induced vibration is suggested. A recent finding by Flamand [37] reveals the dependence of excitation on the thickness of a thin water film moving on the cable surface, and the link between the thickness and surface speed of the water.

There are many other studies on this topic. Yamaguchi's work [32] indicates that this instability is essentially a galloping. Geurts et al. [38] developed a numerical model based on SDOF galloping theory to predict the rain–wind induced cable response for the Erasmus Bridge. Thus obtained analytical results of vibration amplitude approximately agreed with the field data. In the model proposed by Xu and Wang [39], the interaction between wind, cable and rivulet was considered. The predicted steady-state response showed that the main characteristics of an inclined cable with moving rivulet, such as velocity-restricted and amplitude-restricted, could be captured. The analytical study by Gu and Lu [40] pointed out the importance of initial rivulet position in generating the cable instability. The "unstable zone" of initial rivulet position and the "dangerous zone" of instantaneous rivulet position were identified for cables of different natural frequencies. Matsumoto on the other hand finds that the air flow component along the cable is the essential cause of this vibration [41–47].

Various kinds of structural and aerodynamic means have been developed to suppress and prevent the vibration. Increase of the system damping by installation of oil dampers [48], hydraulic dampers [38], viscous-friction dampers [49], or connecting some of the longest cables by using cross-ties [6, 50–52] are found to be effective structural means. The installation of TMD on the vertical hanger of an arch bridge [53] was also proven to be effective. As for developing the aerodynamic measures, the main idea is to break up the formation of the upper water rivulet. Thus, various methods have been proposed and applied to the bridge cables, such as the use of helical wire whirling on the cable surface [34, 54], the adoption of a dimpled cable surface [6, 55], or the use of an axially protuberated surface [56]. All these have proven to be effective and successful in the field to various extents. The effect of cable surface roughness has been also investigated by the wind tunnel tests [57]. A more recent proposal of deflecting the water on cable surface to control the motion needs further development [36].

Regarding the amount of damping required to control it, Irwin [58] has suggested that the vibration can be reduced to a harmless level if the Scruton number is greater than 10, i.e.,

$$Sc = \frac{m\varsigma}{\rho d^2} > 10 \tag{10.3}$$

where, $m =$ cable mass per unit length, $\varsigma =$ the critical damping ratio, $\rho =$ air density, and $d =$ cable diameter. This statement is also supported by experience in Japan [59]. Considering $d = 15$–20 cm and $m = 100$ kg/m for a bridge stay cable, for example, this is equivalent to the structural damping of approximately $\varsigma = 0.5\%$ or more.

10.2.8 High-Speed Vortex Excitation of Dry, Inclined Cables

Although majority of the observed stay cable vibration belongs to the rain–wind induced type, it has been found both in field and wind tunnel tests that dry inclined cables can also undergo large amplitude oscillation without precipitation. Matsumoto et al. [60] reported the observations of cable vibrations without rain but with characteristics of rain–wind vibrations, up to the maximum amplitude of 23 cm at the wind speed 40 m/s during a typhoon. Matsumoto et al. [42] further explained that these vibrations occurred for the cables of the Higashi-Kobe Bridge, too. These phenomena were not properly explained by any known mechanisms and were attributed to the high-speed vortex shedding excitation because, compared to the normal Kármán vortex-induced vibration, the observed instability occurred at much higher reduced wind speed ranges in multiples of 20, i.e., $U_R = 20, 40, 60$, etc.

10.2.9 Dry Inclined Cable Galloping

10.2.9.1 State-of-the-Art

Galloping of dry inclined cables is a relatively new term and its concept has become clearer only recently. First it appeared as a side product of the study on rain–wind vibration. There have been some experimental studies carried out [45–47, 56, 57, 61–64] to investigate this phenomenon and, so far, it has been clearly identified only in wind tunnel tests. Results obtained from these studies show that if the wind is in an oblique direction against the cable plane, the cable could have similar response characteristics as galloping instability. The results are found very sensitive to the model supporting conditions in the wind tunnel tests and thus sometimes it is difficult even to reproduce the same test results. However, if it takes place in reality as predicted, it would be a very serious engineering problem. Although this type of cable motion has never been clearly observed in real bridges, some field observations in fact are said to be better explained with galloping than calling them rain–wind induced vibration [6, 58].

An instability criterion to indicate the critical wind speed originally suggested by [56] for this instability was approximately given by

$$(U_R)_{cr} \approx 35\sqrt{Sc} \tag{10.4}$$

where $U_R = \bar{U}/(fd)$ is the reduced wind speed, and $Sc = m\varsigma/(\rho d^2)$ is the Scruton number. According to Saito [56], this criterion is applicable to the cases where the angle between cable axis and wind direction is 30–60°. It imposed a difficult design condition for bridge stay cables with a typical diameter of 150–200 mm, since it would place so many bridge stay cables into a category of "prone to galloping".

The reality, however, is that many existing stay cables seem to be surviving without suffering this instability. Further investigation was thus urgently required under these circumstances.

Wind tunnel investigation on inclined cable vibration has been attempted by many researchers since then and extensive results have been made available. The state-of-of-the-art of this particular instability is that

1. Galloping of dry inclined cable does exist as a possible instability.
2. It is a unique aerodynamic phenomenon for a cable, which is inclined against wind.
3. Instability takes place in the critical Reynolds number range.
4. There are a few specific geometrical positions where the cable could become unstable.
5. The criterion given by (10.4) seems to be too conservative. It probably includes some of the high-speed vortex excitation cases.

10.2.9.2 Critical Issues

Mechanism

There are some interesting issues raised in relation to the excitation mechanism of this instability. One of them is the role of the air flow along the cable axis and possible interaction of it with the Kármán vortices behind the cable. It has been suggested by Matsumoto et al. [47], who also indicate that the axial flow in fact has a significant role in rain–wind vibration, too, as mentioned earlier in Sect. 10.2.7. By introducing an artificial axial flow in the experiment, Matsumoto et al. [47] showed that its existence could induce negative slope of the lift force, and galloping of dry inclined cable would occur when the velocity of the axial flow was 30% or more than that of the approaching flow. A question then is whether or not this artificially imposed axial flow well represents the air flow situation behind an inclined cable in reality.

Critical Reynolds Number

Another significant fact is, as pointed out by Larose and Zan [65], that the instability is observed clearly in the critical Reynolds number range. A question is then why and how it relates to the instability. It is also extremely interesting to note that the conventional Den Hartog criterion would indicate instability corresponding to the drag crisis in this range, but by considering the large influence of Reynolds number on force coefficients. The governing reason of inclined cable galloping is actually found to be the difference in lift force due to Re, the $\partial C_L/\partial \mathrm{Re}$ term [66].

The critical Reynolds number is sensitive to the presence of surface roughness, flow turbulence, motion of the cable and a flow angle not perpendicular to the cylinder axis. More recent study further indicate the fact that the effects of Reynolds number on aerodynamic force distribution and resulted cable motion are highly dependent upon the orientation of the body to the mean flow direction [67–69].

Onset Criteria

Macdonald [69] has tried a general quasi-steady analysis of inclined cable galloping, including the effect of Reynolds number variation, and derived a general 2D criterion for galloping instability given as follows:

$$Z_s = \frac{\varsigma_s m f_n}{v/\rho} > R\left[\frac{\mathrm{Re}}{16\pi}\left\{-h(C_D) + \sqrt{g^2(C_D) + g^2(C_L) - h^2(C_L)}\right\}\right] \quad (10.5)$$

where $\rho =$ the air density, $\mathrm{Re} = Ud/v =$ Reynolds number, $d =$ cable diameter, $C_D =$ drag coefficient, $C_L =$ lift coefficient, $\omega_n = 2\pi f_n =$ circular natural frequency, $\varsigma_s =$ the structural damping ratio, $m =$ cable mass per unit length, and $U =$ wind speed. $R[\ldots]$ indicates the real part. Also

$$g(C_F) = C_F\left(2\sin\phi - \frac{1}{\sin\phi}\right) + \frac{\partial C_F}{\partial \mathrm{Re}}\mathrm{Re}\cdot\sin\phi + \frac{\partial C_F}{\partial\phi}\cos\phi \quad (10.6a)$$

$$h(C_F) = g(C_F) + \frac{2C_F}{\sin\phi} \quad (10.6b)$$

in which C_F is either C_D or C_L and $\phi =$ the cable-wind angle, defined as the angle between a flow velocity relative to the cylinder axis.

It is important to note that the right hand side of the criterion (10.5) is a function of Re and ϕ only and independent of the direction of cable motion. Note also that the expression (10.5) is simplified for a circular cylinder. More general expression and its derivation are given in Macdonald and Larose [66]. Detailed measurement of both lift and drag force components for an inclined cylinder in the critical Reynolds number range was carried out by Larose et al. [67, 68]. It is interesting to note that Macdonald has predicted, based on this analysis and available aerodynamic data, that another instability area, which exists when the cable-wind angle is between 75° and 90°. A practical importance of this analysis is in the fact that the magnitude of aerodynamic negative damping can be actually calculated, and hence, the additional damping to the system to suppress any instability can be predicted.

10.3 Field Observation of Wind-Induced Bridge Cable Vibrations

There are a number of reports that cable-stayed bridges suffered cable vibration induced by wind. Though in many cases, some of the physical parameters, such as the exact wind speed and direction relative to the cable, are not fully documented in detail, these reports from the field are no doubt the most important information for any engineering solution required for the problem. Table 10.1 is a list of some of the incidents immediately known to us with the sources of information. As it is evident from the list, the majority of known cable vibrations are attributed to the rain–wind vibration.

It should be added here that there are also reports regarding the vibration of cables other than bridge stay-cables, including the power transmission lines [81], steel hangers of an arch bridge [53], and inclined hangers of a long-span suspension bridge [82], all of which were caused by combined action of rain and wind.

Rain–wind vibration was clearly identified as a new type of cable vibration at the Meiko-Nishi Bridge in 1984. Hence, the two preceding cases of Köhlbrand and Brotonne in the above table were considered to belong to this category in later days.

By reviewing the cable behaviour in above cases, the commonly observed characteristics of rain–wind vibration can be summarized as following [27, 41, 44, 48, 51, 83]:

1. Moderate rain—neither light drizzle nor a downpour is conductive to such vibrations
2. Wind speed of 6–18 m/s, with the majority of the cases in 8–12 m/s
3. Cable frequency of 0.6–3.0 Hz
4. Cable diameters ranging from 140 to 225 mm
5. Reynolds number of 6×10^4–2×10^5, which is the transition range from sub-critical to critical
6. Vibrating cables are located in the leeward side of the pylons, most of the case
7. Cable inclination of 20–45° from horizontal, in many cases
8. Wind direction of 20–60° relative to the plane of the cable

Not too many cases are known where the wind induced vibrations other than rain–wind excitation were witnessed. Table 10.2 is a summary of recently known cases.

Macdonald [90] derived an expression for the aerodynamic damping of an inclined cable as follows:

$$\varsigma_a = \frac{\rho U d C_D}{4\omega_n m}\left\{\sqrt{\cos^2\theta + \sin^2\theta \sin^2\beta} + \frac{\sin^2\theta \sin^2\beta}{\sqrt{\cos^2\theta + \sin^2\theta \sin^2\beta}}\right\} \quad (10.7)$$

where $\rho =$ the air density, $d =$ cable diameter, $C_D =$ drag coefficient, $\omega_n = 2\pi f_n =$ circular natural frequency, $m =$ cable mass per unit length, $U =$ wind speed, $\theta =$ wind direction measured from normal to the vertical plane of the

Table 10.2 Observed vibrations of other types

Name	Country	Year	References
Galloping with snow/sleet			
Ørsund	Denmark/Sweden	2004	Larsen and Lafrenière [12]
Wake galloping			
Yobuko	Japan	1989	Narita and Yokoyama [84], Yoshimura et al. [74]
Akashi	Japan	1998	Toriumi et al. [21]
Vortex excitation			
Evripos	Greece	?	Virlogeux [6]
Fred Hartman	Texas, USA	?	Zuo et al. [85]
Gjermnessunder	Norway	?	Hjorth-Hansen and Strømmen [86]
Parametric excitation			
Wandre[a]	Belgium	1988	Lilien and Pinto da Costa [75]
Helgeland	Norway	1993	Asa-Jakobsen et al. [87]
Guazú	Argentina	1996	Andersen et al. [88]
Second Seven	UK	1990s	Irwin [80], Macdonald et al. [90]
Fred Hartman	Texas, USA	?	Liu et al. [26]
Guadiana	Portugal	2003	Caetano et al. [89]

[a]For this incident, Cremer et al. [76] claim it as the rain–wind vibration, whereas Lilien and Pinto da Costa [75] say "there was no rain" and it was induced by motion of the cable end

cable, and $\beta =$ angle of inclination of the cable from the horizontal. With the help of this expression and a brief summary of the past record, he further discussed a benchmark damping that would be necessary in order to suppress these wind-induced vibrations in terms of the Scruton number. He indicated as follows:

Rain–wind vibration	$Sc > 10$ by Irwin [80] seems reasonable.
Vortex excitation	$Sc > 5$ as suggested by Scruton and Walshe [91]. Small amplitude anyway.
Galloping	There is not enough evidence to establish a clear criterion.
Parametric excitation	Difficult to control by simply increasing damping. Further clarification required.

10.4 Vibration Control in Practice

10.4.1 General

Civil engineering structures including bridges, buildings and towers usually have the overall structural damping at the level of 1% of critical. Hence, if there is any possibility of serious dynamic excitations anticipated and yet damping is significantly lower than this level, there should be a warning flag. Cables belong to this category almost always and this is why cable vibrations require a particular engineering attention. They are very flexible, often exposed to fluid dynamic excitations and its inherent damping is usually the order of magnitude less than

Table 10.3 Observed structural damping of cables

Macdonald [92]	Experience at the 2nd Severn 0.047–0.33%. Higher for shorter cables
Stubler et al. [27]	0.08–0.13% for grouted parallel strand; 0.15–0.30% for locked coils
Virlogeux [6]	0.01–0.20%
Yamaguchi and Fujino [93]	0.16% for the first mode and lower for higher modes

other structural members. Past experience on structural damping of cables indicate a wide range of values as in Table 10.3.

Introduction of artificial dampers is naturally an attractive consequence. There have been various dampers developed and applied to power transmission lines and bridge cables alike. Some of them are referred here.

10.4.2 Dampers Commonly Used for Power Lines

The present document is particularly regarding the aerodynamics of cables employed for bridges. However, it is worth reviewing the past experience in the field of power line vibration since there have been so many types of relatively inexpensive, simple dampers developed for power line applications. There is a comprehensive summary of them available in existing literature ([94], for example). Aeolian vibration dampers widely used are typically as follows:

The *Stockbridge type damper* consists of two pear-shaped end-masses attached to a steel strand, which is often referred to as the messenger cable, and the middle of the messenger cable is clamped and fixed to the conductor. When the end-masses vibrate in their natural frequencies, the messenger cable is bent and friction caused by slipping between wires dissipates vibration energy. Two masses and the strands are sometimes made non-symmetric to make the damper more sensitive to different frequencies. A possible problem of this damper is fatigue failure of the messenger cable, ironically particularly when the damper is effectively working. It was one of the earliest damping devices, and was already commercially available in 1924. However, it went through a series of development and modifications since then. There are also various size and models of it in different countries. The Stockbridge damper has been used for bridge cables, too.

Torsional damper was invented in Canada and is effective particularly when the conductor vibrates in twisting mode. It consists of two masses, one above and one below the conductor, connected by an arm which is bolted to the conductor through a polymer disc. When twisting vibration of the conductor occurs, the polymer is shear loaded due to the inertia of the damper masses.

Impact damper is a Swedish invention named Elgra. It consists of a vertical stem cramped on the conductor and three cast masses loosely fitted to this shaft. Each mass rests on a polymer washer. When the conductor acceleration exceeds 1 g (=9.81 m/s^2), these masses do not follow the conductor motion any more and start rattling, which results vertical impacts and dissipates energy. Field observations have indicated that it actually performs well when the acceleration is beyond 2 g [94].

This type of damper can be developed more and probably used for cable-supported civil engineering structures, too. Possible problems need to be addressed in future are noise and possibility of falling masses in case they are broken.

The *bretelle damper* is not a commercially available damper. It is simply a jumper loop of about 2–5 m long, made of a scrap conductor piece, connecting the adjacent spans at both sides of a support. It seems to work as a damper by transferring vibration energy between the spans. It has been widely used in France. It can be regarded as a type of secondary cable systems, which can have a wide variety of applications.

There are a number of other types of aeolian vibration dampers mentioned in the literature [94] together with the past experience and experimental observations. In comparison to the vortex shedding excitation case, wake-induced oscillations of bundled conductors are often controlled by tilting the bundles, adjusting the cable separation and/or staggering the sub-span systems. Associated with the last method, flexible *spacer-dampers* have been used but they are often structurally complicated and expensive.

More serious vibration problems for power-lines are galloping and vibrations related to wake interference. However, most of the damping mechanisms or protection measures used to prevent these vibrations and their damages are not directly applicable to cable vibrations in general.

10.4.3 Dampers Used for Bridge Cables

There have been various attempts to control vibrations of stay cables of cable-stayed bridges and hanger ropes of suspension bridges. Table 10.4 summarizes Japanese experience, where vibration control of bridge cables was attempted. The simplest of them is the use of *auxiliary ropes* in various ways. They can be installed in horizontal, vertical or diagonal ways depending on the cases but the fundamental idea is the same. The method has been applied to many bridges including Farø, Normandy, Meiko-Nishi, Hitsuishi, Iwaguro etc. Figure 10.3 is an example of auxiliary wires for the case of the Yobuko Bridge, Japan.

A little more sophisticated method but similar to it is the use of *spacer-dampers* to provide the connectors or spacers with a simple damping mechanism. The Normandy Bridge and the Yokohama-Bay Bridge are equipped with them. The idea can be extended to placing an auxiliary cable networks even with dampers integrated into it. Some analytical and experimental attempts have been made on in recent years.

Many stay cables of cable-stayed bridges are equipped with a simple *visco-elastic anchorage* system at their connection to the bridge deck. This can be as simple as having a neoprene rubber washer inside the anchoring socket. If the damping by anchorage structure is not enough, a shock absorber or a *dashpot* can be installed at near the anchorage. Brotonne, Sunshine Skyway, Aratsu and some other bridges have these dampers and they seem to be quite effective. In order to make the installation of the damper most effective, the damper's damping can be properly designed by considering the overall modal damping. Pacheco et al. [95] obtained a

Table 10.4 Japanese experience of vibration control for bridge cables

Bridge									Countermeasure							
Style					Cable (Hanger)				Device							
Name	Steel	PC	Com	Span (m)	Year	*L* (m)	*d* (mm)	Arrange	OD	VSD	HDR	Wire	Sp	Aero	Phenomenon	Remarks
Cable-stayed bridge																
Rokko Bridge	○			90 + 220 + 90	1977	53–103	85.8	S				○			G	Hexagonal cable
Meiko-Nishi Bridge	○			176.5 + 405 + 176.5	1985	65–195	125–165	S			○	●			K,RV	Wire was adopted in the beginning of opening
Aratsu Bridge	○			185 + 115 + 45	1988	50–166	160,180	S	○						K,RV	
Hitsuishi-Iwakuro Bridge	○			185 + 420 + 185	1988	70–192	147–187	D				○	○		WG,RV	
Yokohama Bay Bridge	○			200 + 460 + 200	1989	75–221	122–175	D	○				○			Damper was installed in spacer
Tempozan Bridge	○			170 + 350 + 120	1989	73–186	160–200	S	○						K,RV	
Shima Maruyama Bridge		○		113.4 + 113.4	1989	16–112	82	○				○				
Yobuko Bridge		○		121 + 250 + 121	1989	16–136	75.5–83.5	D				○			WG	
Sakitama Bridge	○			190 + 190	1991	about 200	142–173	D		○					K,R,V,WG	
Tomei Ashigara Bridge		○		92 + 185 + 92	1991	19–100	140–160	S							RV	
Ikuchi Bridge			○	150 + 490 + 150	1991	54–246	110–139	S			Δ				K	Rubber in for reducing additional stress
Aomori Bay Bridge		○		128 + 240 + 128	1992	48.2–157.5	212	D		○					RV	
Usui Bridge		○		111 + 111	1992	45–118	7ϕ163–241	D		○					B	
Higashi-Kobe Bridge	○			200 + 485 + 200	1993	66–220	145–155	S						○		Cable with axial protuberances
Sun Marine Bridge	○			60 + 145 + 45	1993	45–115	110–155	S		○					RV	
Fukiagehama Sunset Bridge				No Info.	1993	No Info				○						
Tsurumi Tsubasa Bridge	○			255 + 510 + 255	1994	84–283	149–190	S	○		○				K,RV	Combined use

Okutama Bridge			○	105 + 160	1994	45–115	7ϕ91–337	S						○		Cable with axial protuberances
Matsukawaura Bridge				No Info.	1994	No Info				○						
Chichibu Park Bridge		○		195 + 195	1994	41.9–207.8	75.5–96	D				○			WG	
Tajiri Sky Bridge		○		168.5 + 168.5	1994	27–183	120–155	S		○					RV	
Ogawa Bridge		○		114.9 + 114.9	1995	34–115	140	S		○					RV	
Shinwatashi Bridge		○		93.75 + 93.75	1995	15.2–101.3	74–82.5	D				○			WG	
Daini Chikumagawa Bridge		○		135 + 135	1996	No Info					○					
Yuge Bridge	○			75 + 175 + 75	1996	24–79	68–110	S			○			○	K, RV	Aero for longest cable, HDR for next three steps
Ikara Bridge		○		120 + 260 + 120	1996	14.4–140.6	55.5–76.0	D		○		○	○		WG	
Shonan Ginga Bridge	○			107 + 242 + 107	1997	39–117	150–225	S			○				K, RV	
Daishi Bridge	○			113 + 203	1997	46–148	105–180	S			○				K, RV	
Fuchuyotsuya Bridge	○			93 + 260 + 93	1997	55–129	No Info	S								
Fureai Bridge				123.1 + 49	1997	23–84	95–173	S			○				K, RV	
Tokachi Bridge		○		124.1 + 251 + 124.1	1997	14–130	200	D	○						RV	Low temperature type oil damper was adopted
Meiko Central Bridge	○			290 + 590 + 290	1998	90–302	136–161	S		○					K, RV	
Shin-Inagawa Bridge		○		200 + 200	1998	62.7–198.5	135,150	S			○				RV	
Meiko Higashi Bridge	○			145 + 410 + 145	1998	61–209	199–349	S			○				K, RV	
Koremasa Bridge	○			121.2 + 186.6 + 91.4	1998	129	119	S			○				K, RV	Conducted only for the longest cable
Ohshima Bridge	○			160 + 350 + 160	1999	No Info					○					
Naka-Noto Bridge		○		109 + 230 + 109	1999	No Info	145–160	S		○					K,RV	

(continued)

Table 10.4 (continued)

Bridge									Countermeasure							
Style					Cable (Hanger)				Device							
Name	Steel	PC	Com	Span (m)	Year	L (m)	d (mm)	Arrange	OD	VSD	HDR	Wire	Sp	Aero	Phenomenon	Remarks
Tatara Bridge			○	270 + 890 + 320	1999	108–462	110–170	S			Δ			○	RV	Aero: Indent type, HDR: Rubber Seal
Cable-stayed bridge																
Tenkenji Bridge		○		102.7 + 219 + 102.7	1999	18–116	165,120	S		○					RV	
Minami-Honmoku Bridge	○			No Info.	2000	No Info					○					
Daini-Satsunai Bridge	○			No Info.	2000	No Info					○					
Miyanoura Bridge		○		No Info.	2000	No Info					○					
Hoshinofurusato Bridge		○		131.7 + 131.7	2000	25–136	166,212	S			○				RV	
Uchinada Bridge		○		86 + 172 + 86	2000	23.4–94.3	115,120	S			○				RV	
Extradosed bridge																
Odawara Blue Way Bridge		○		74 +122 + 74	1994	25.7–51.5	125	S			○				RV	
Tsukuhara Bridge		○		65.4 + 180 + 76.4	1998	No Info					○					
Cable-stayed bridge with Arch-type tower																
Haneda Sky Arch	○			80 + 80	1993	No Info		○		○					KV	
Suspension bridge																
Akashi-Kaikyo Bridge	○			960 + 1990 + 960	1998	max 208	85–91	D						○	KV, WIF	
Kurushima Bridge	○			280 + 1030 + 260	1999	max 104	72–80	S			○				K	Seal material was used
Akinada Bridge	○			255 + 750 + 170	2000	max 78	69–103	S			○				K	

(*Remarks*): Com: composite bridge
Arrangement: *S*: single, *D*: double, *O*: other
Countermesures: *OD*: oil damper, *VSD*: viscous shear damper, *HDR*: high-damping rubber type damper
Wire: cable cross ties, *Sp*: spacer, *Aero*: aerodynamic countermeasure
Phenomenon: *K*: kármán vortex-induced vibration, *RV*: rain/wind-induced vibration, *WG*: wake galloping, *B*: buffeting (blank shows no information)
○: adopted, ●: adopted in the past (now other countermeasures are adopted)
Δ: (HDR) adopted, but the primary purpose is different (reducing additional stress by live load or waterproofing)

Fig. 10.3 Auxiliary wire system of the Yobuko Bridge

Fig. 10.4 Oil dampers installed on the Aratsu Bridge

universal design curve based on general physical parameters considered for a dashpot damper. Krenk and Høgsberg [96] further analysed the system and came up with a general asymptotic format to determine effective damping from a few parameters. Recent studies [97, 98] include effects of various parameters on the estimation of modal damping of stay cables and accuracy of the estimation is greatly improved. Figures 10.4–10.8 show application of these dampers to some of the Japanese bridges. Figures 10.4 and 10.5 show dashpot type dampers applied

Fig. 10.5 Viscous shear dampers of the Sakitama Bridge

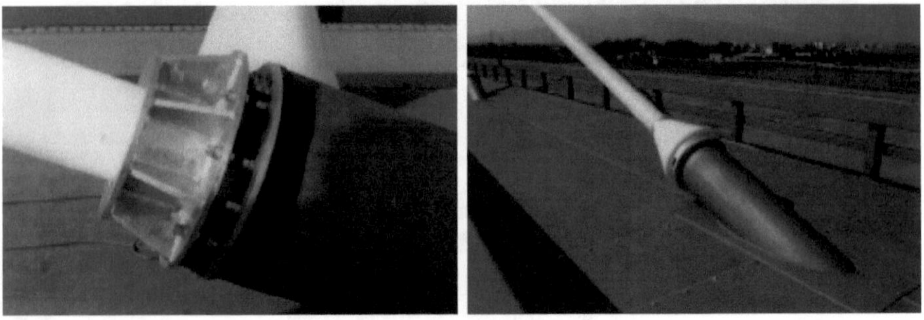

Fig. 10.6 High-damping rubber damper of the Shonan Ginza Bridge

to Aratsu and Sakitama. Figure 10.6 shows the application of high-damping rubber, which was the case of the Shonan Ginza Bridge, whereas Fig. 10.7 is an example in which both high-damping rubber and oil dampers are used together. A similar damper can be applied to the vertical hanger ropes, as the case of the Kurushima suspension bridges shown in Fig. 10.8.

Newer examples of more sophisticated dampers include the *friction damper* developed by VSL [99] and the *magneto-rheological fluid damper* [100]. The former was applied to the Uddevalla Bridge, Sweden, and the latter is operated at the Eiland Bridge, the Netherlands. Both of them are reported to be quite effective.

The VSL friction damper [99] has a relatively simple structure: it consists of two parts. There are steel wings in a plane normal to the cable and attached on the cable by a collar. Pressed on them are the spring blades, which are rigidly fixed on the bridge deck. As a result, when the steel wings move with the cable, they will dissipate energy by frictional contact with the spring blade.

Fig. 10.7 Hybrid use of high-damping rubber and oil dampers for the Tsutsumi Tsubasa Bridge

Fig. 10.8 A damper for a hanger rope used for the Kurushima-Kaikyo Bridges

Actually the majority of stay cable vibrations are caused by rain–wind excitation mechanism and they are often effectively controlled by giving various *surface treatment* to the stay cables rather than increasing damping. Giving axial grooves, helical strakes, indentation roughness and attaching semi-circular fins are some examples. Shown in Fig. 10.9 is the cross-section of the Higashi-Kobe Bridge cable with longitudinal fins, whereas Fig. 10.10 shows the Tatara Bridge cable with surface indentations.

Fig. 10.9 Parallel protuberances on the cable surface at the Higashi-Kobe Bridge

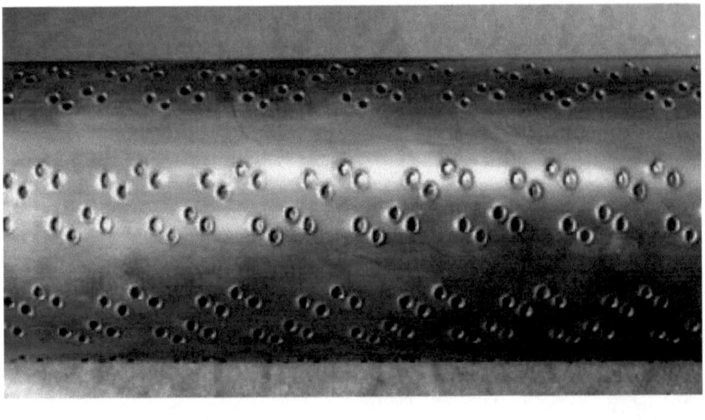

Fig. 10.10 Indented cable surface at the Tatara Bridge

The *Stockbridge dampers* have been used for suspension bridge hangers, too, including Severn, Humber and Bosporus.

10.4.4 Cases of Vibration Control in Japan

10.4.4.1 The Tempozan Bridge

The Tempozan Bridge is a cable-stayed bridge with a 350 m steel span and double-plane stay-cables. Provoked by preceding examples of Meiko-Nishi and Akashi, a

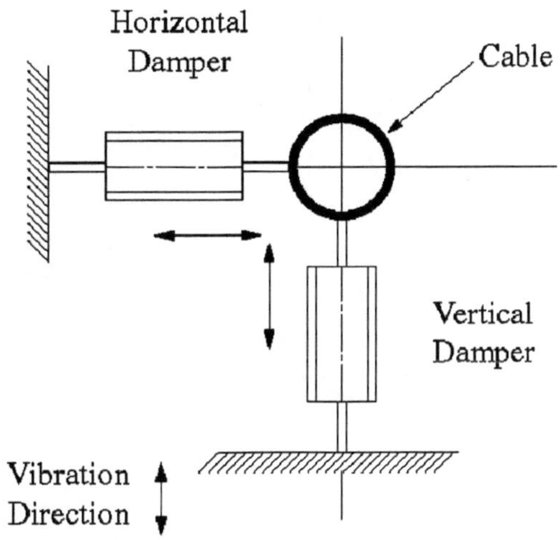

Fig. 10.11 Damper arrangement at the Tempozan Bridge

series of wind tunnel tests was carried out and a possibility of the rain–wind vibration was indicated. As the control methods, the auxiliary wires, both fiber and steel ropes, were tried first particularly aiming at the erection period. For a period of 18 months after its installation, only the higher mode vibration with very small amplitude was sometimes observed, having the nodal points at the joints of the auxiliary wires and the stay cables [101, 102].

Finally, as a more permanent measure, the oil dampers were installed inside the bridge fairing considering their effectiveness and also the aesthetic view. This design resulted in a very short distance between the position of damper and the end of the cable, the effectiveness of the damper's performance under such condition was experimentally confirmed prior to its installation. A set of two dampers, as a result, were arranged in perpendicular ways, as shown in Fig. 10.11, since the rain–wind vibration was found to be predominantly in the in-plane or vertical direction but also with an out-of-plane or horizontal motion component.

The required additional damping to suppress motion was found to be $\delta = 0.04$, or 0.64% of critical. No conspicuous cable vibration has been reported on this bridge since the installation of dampers.

10.4.4.2 The Tatara Bridge

With its 890 m centre span, the Tatara Bridge recorded the world longest crossing as a cable-stayed bridge at its opening. The Tatara cables are very long and flexible and the natural frequencies are extremely low. Prior to the bridge construction, the dynamic model of the stay-cable was wind tunnel tested using a steel cable model, which had a 155 mm diameter, 12 m long and covered with a PE tube. From the tests, the rain–wind vibration was anticipated in the long stay-cable, whose frequencies were 0.26–0.54 Hz. It was found that the vibration amplitude would

Fig. 10.12 Elastic sealing material applied to the stay-cable of the Tatara Bridge

be restricted to a harmless range, if the damping is as high as $\delta = 0.02$, or 0.32% of critical [103].

Adoption of the secondary wire systems was found to be problematic because of possible fatigue damage of wires, durability of connections as well as aesthetics. Also, if the viscous dampers are to be installed, they need to be located at rather high positions from the deck surface in order to obtain effectively high damping. Treatment of the cable surface by applying parallel protuberances, as adopted for the Higashi-Kobe Bridge, is known to be effective but at the expense of significant increase in drag force on the cables. The new solution adopted for the present bridge was to apply random indentations on the cable surface as shown in Fig. 10.10, which effectively removed the rain–wind vibration. Against the vortex-induced vibration, which was often observed during the bridge erection, the elastic sealing materials were tried out at the site and found to be effective (Fig. 10.12). In addition, the high-damping rubber dampers were installed at the mid-span to augment the overall damping of the structure.

10.4.4.3 The Akashi-Kaikyo Bridge

The maximum length of hanger ropes of the Akashi Kaikyo Bridge exceeds 200 m. PWS hangers with PE surface coverage were adopted for this bridge because of their structural reliability and relatively easy efforts required for installation and maintenance. Hangers are installed as a set of two parallel ropes in a plane perpendicular to the bridge axis. The clearance between two hanger ropes is nine time the hanger diameter and both ends of each set of hangers are pin-jointed as shown in Fig. 10.13 [21, 104].

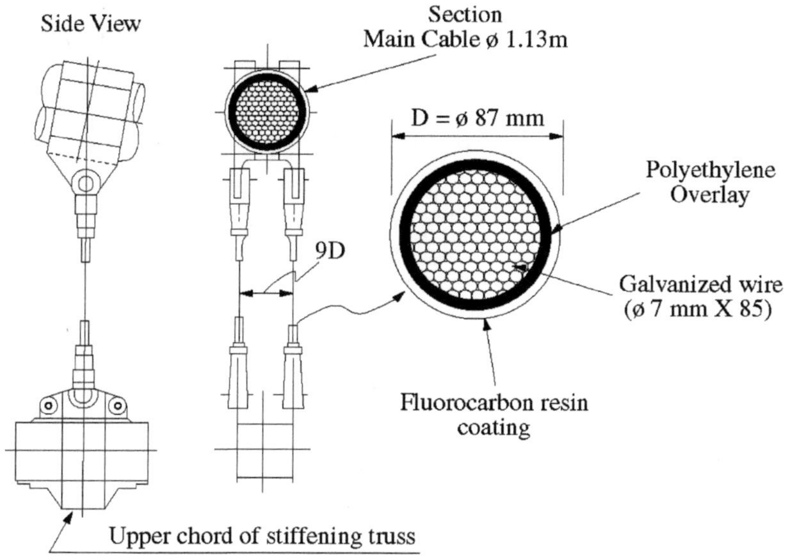

Fig. 10.13 Hanger ropes of the Akashi Kaikyo Bridge

At the bridge erection stage, two types of aerodynamic excitation of these hangers were observed. One was clearly the vortex-induced vibration and the high-damping rubber dampers were installed to successfully reduce its magnitude. Another type of vibration was apparently due to wake-induced excitation, which was observed only in the leeward cables of each parallel set. This vibration was observed not too often, but it was a large amplitude motion. The aerodynamic characteristics for the case of tandem cylinders with 9*d* interval are usually deemed stable and so it was not investigated at the design stage. However, the hangers did vibrate for this case.

Careful wind tunnel tests using a full-scale sectional model of the parallel hangers revealed that, with this configuration, the downstream hanger was expected to experience the negative aerodynamic damping of $\delta = -0.22$ and it would vibrate in an elliptic orbit with its longer axis along the bridge. These findings agreed well with the field observations.

In order to reduce the aerodynamic interaction between two parallel hangers, the installation of helical stripes around the hangers was found useful. It was found most effective to have 9 mm diameter ropes in helical configuration with 800 mm pitch, as shown in Fig. 10.14, to suppress both the wake-induced instability and vortex-shedding excitation. The helical ropes were installed on 45 long hanger sets from each tower, resulting in a total of 360 hanger sets for the whole bridge.

Actually, against the vortex-shedding excitation, there were dampers installed on these hangers before this problem took place. However, all of these dampers were removed later because most of them were damaged by the wake-induced instability during the passage of the typhoon Vicki (9807). Also, the helical stripes were effective enough for suppressing the vortex-induced vibration as well.

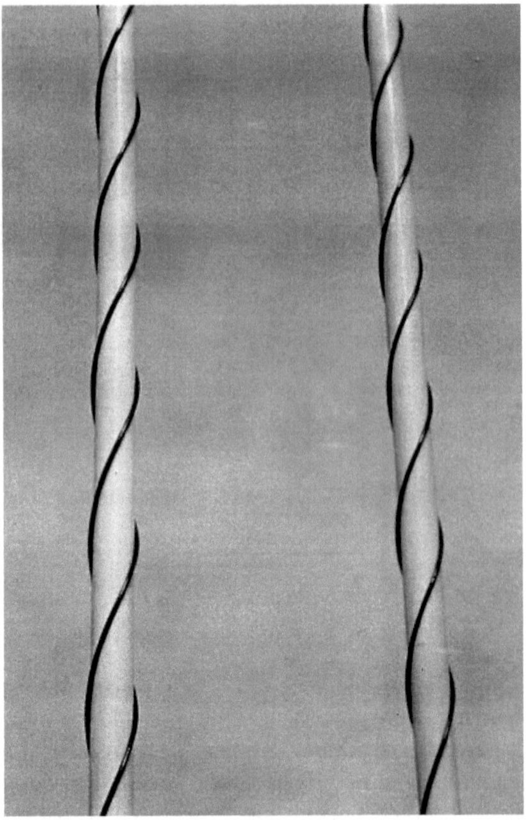

Fig. 10.14 Helical stripes installed on the hangers of the Akashi-Kaikyo Bridge

The selected material for the helical ropes was wrapped fiber ropes with 10 mm diameter, considering the shrinkage of the cross-section due to the stretch of ropes during their installation. The installation of helical ropes was done in May through September of 1999. After the installation, the wake-induced instability was never observed, despite the fact that the bridge site was exposed to two major typhoons, Zia (9916) and Bart (9918).

Acknowledgments Some parts of this chapter, namely 10.1–10.2.8, are primarily based on the chapter 12 of reference [105], which was originally prepared by Hiroshi Tanaka and Helmut Wenzel. Earlier discussion on the same material by Guy L. Larose and Shaohong Cheng is also acknowledged.

References

1. Ehsan F, Scanlan RH, Bosch HR (1990) Modeling spanwise correction effects in the vortex-induced response of flexible bridges. J Wind Eng Ind Aerod 36(2):1105–1114
2. Leblond A, Hardy C (1999) On the estimation of a $\theta \leq 45°$complex matrix of symmetric Stockbridge-type dampers. In: Proceedings of the 3rd international symposium on cable dynamics (ISCD), Trondheim, pp 139–144

3. Sauter D, Liu Q, Hagedorn P, Rahman, A (2001) On the vortex-excited vibrations of stay-cables. In: Proceedings of the 4th international symposium on cable dynamics (ISCD), Montreal, pp 277–284
4. ASCE Committee on Electrical Transmission Structures (1984) Guidelines for transmission line structural loading, ASCE Structural Division
5. Davenport AG (1979) Gust response factors for transmission line loading. In: Proceedings of the 5th international conference on wind engineering, Fort Collins, Colorado, pp 899–910
6. Virlogeux M (1998) Cable vibrations in cable-stayed bridges. In: Larsen A, Esdahl S (eds) Bridge aerodynamics. Balkema, Rotterdam, pp 213–233
7. Den Hartog JP (1932) Transmission line vibration due to sleet. AIEE Trans 51:1074–1076
8. Lilien JL (1997) Galloping of overhead electrical lines, mechanisms, wind tunnel experiments—field measurements. In: Proceedings of the 2nd international seminar on cable dynamics (ISCD), Tokyo, pp 37–48
9. Novak M, Tanaka H (1974) Effect of turbulence on galloping instability. Proc ASCE 100 (EM1):27–47
10. Ito M, Fujino Y, Yamaguchi H (1985) Wind tunnel study on galloping of suspended figure-8 telecommunication cables. Proc JSCE Struct Eng 2(1):217–225
11. Novak M, Davenport AG, Tanaka H (1978) Vibrations of towers due to galloping of iced guy cables. Proceedings ASCE 104(EM2):457–473
12. Larsen A, Lafrenière A (2005) Application of a limit cycle oscillator model to bridge cable galloping. In: Proceedings of the 6th international seminar on cable dynamics (ISCD), Charleston, South Carolina
13. Zdravkovich MM (1997) Flow around circular cylinders. Oxford University Press, Oxford
14. Yoshimura T, Savage M, Tanaka H (1995) Wind-induced vibrations of bridge stay-cables. In: Proceedings of the 1st international seminar on cable dynamics, Liège, pp 437–444
15. Chen SS (1987) A general theory for dynamic instability of tube arrays in cross flow. J Fluid Struct 1:35–53
16. Wardlaw RL (1994) Interference and proximity effects. In: Sockel H (ed) Wind-excited vibrations of structures, Springer-Verlag, pp 321–363
17. Wardlaw RL, Cooper KR (1974) Mechanisms and alleviation of wind-induced structural vibrations. In: Proceedings of the 2nd symposium on applications of solid mechanics, Hamilton, Ontario, pp 369–399
18. Simpson A, Flowers JW (1977) An improved mathematical model for the aerodynamic forces on tandem cylinders in motion with aeroelastic applications. J Sound Vib 51(2):183–217
19. Hardy C, Bourdon P (1979) The influence of spacer dynamic properties in the control of bundle conductor motion. IEEE PES summer meeting, Vancouver
20. Price SJ, Paidoussis MP (1984) The aerodynamic forces acting on groups of two and three circular cylinders when subject to a cross-flow. J Wind Eng Ind Aerod 17:329–347
21. Toriumi R, Furuya N, Takeguchi M, Miyazaki M, Saito Z (1999) A study on wind-induced vibration of parallel suspenders observed at the Akashi-Kaikyo bridge. In: Proceedings of the 3rd international symposium on cable dynamics, Trondheim, pp 177–182
22. Kovács I, Leonhardt A (1982) Zur frage der Seilschwingungen und der Seildämpfung. Die Bautechnik 10:325–332
23. Fujino Y, Warnitchai P, Pacheco B (1993) An experimental and analytical study of auto-parametric resonance in a 3DOF model of cable-stayed-beam. J Nonlinear Dynam 4:111–138
24. Warnitchai P, Fujino Y, Susumpow T (1995) A non-linear dynamic model for cables and its application to a cable-structure system. J Sound Vib 187(4):695–712
25. Gani F, Tanaka H (2005) Parametric excitation of stay-cables. In: Proceedings of the 6th ISCD, Charleston, South Carolina
26. Liu MY, Zuo D, Jones NP (2005) Deck-induced stay cable vibrations: field observations and analytical model. In: Proceedings of the 6th ISCD, Charleston, South Carolina
27. Stubler J, Ladret P, Domage JB, Peltier M (1999) Bridge stay-cable vibration: phenomena, criteria and damper technology. In: Proceedings of the 3rd ISCD, Trondheim, pp 163–170

28. Davis DA, Richards DJW, Scriven RA (1963) Investigation of conductor oscillation on the 275 KV crossing over the River Severn and Wye. Proc IEE 110:205–219
29. Richards DJW (1963) Aerodynamic properties of the Severn Crossing conductor, In: Proceedings of the 1st ICWE, Teddington 2:688–765
30. Hikami Y (1986) Rain vibrations of cables of cable stayed bridge (in Japanese). J Wind Eng JAWE 27:17–28
31. Yamada Y, Shiraishi N, Toki K, Matsumoto M, Matsuhashi K, Kitazawa M, Ishizaki H (1991) Earthquake-resistant and wind-resistant design of the Higashi-Kobe bridge. In: Ito M, Fujino Y, Miyata T, Narita N (eds) Cable-stayed bridges: recent development and their future, Elsevier, Amsterdam 397–416
32. Yamaguchi H (1990) Analytical study on growth mechanism of rain vibration of cables. J Wind Eng Ind Aerod 33:73–80
33. Ruscheweyh H (1999) The mechanism of rain–wind-induced vibration. In: Larsen A, Larose GL, Livesey FM (eds.) In: Proceedings of the 10th international conference on wind engineering: wind engineering into the 21st century. Balkema, Copenhagen, pp 1041–1047
34. Bosdogianni A, Olivari D (1996) Wind- and rain-induced oscillations of cables of stayed bridges. J Wind Eng Ind Aerod 64:171–185
35. Verwiebe C (1998) Rain–wind-induced vibrations of cables and bars. In: Larsen A, Esdahl S (eds) Bridge aerodynamics. Balkema, Rotterdam, pp 255–263
36. Verwiebe C, Ruscheweyh H (1998) Recent research results concerning the excitation mechanisms of rain–wind-induced vibrations. J Wind Eng Ind Aerod 74–76:1005–1013
37. Flamand O (2001) An explanation of the rain–wind induced vibration of inclined stays. In: Proceedings of the 4th ISCD, Montreal, pp 69–76
38. Geurts C, Staalduinen P (1999) Estimation of the effects of rain–wind induced vibration in the design stage of inclined stay cables. In: Larsen A, Larose GL, Livesey FM (eds) Wind engineering into the 21st century. Balkema, Rotterdam, pp 885–892
39. Xu YL, Wang LY (2001) Analytical study of wind-rain-induced cable vibration. In: Proceedings of the 5th Asia-Pacific conference on wind engineering, Kyoto, pp 109–112
40. Gu M, Lu Q (2001) Theoretical analysis of wind–rain induced vibration of cables of cable-stayed bridges. In: Proceedings of the 5th Asia-Pacific conference on wind engineering, Kyoto, pp 125–128
41. Matsumoto M (1998) Observed behaviour of prototype cable vibration and its generation mechanism. In: Larsen A, Esdahl S (eds) Bridge aerodynamics. Balkema, Rotterdam, pp 189–211
42. Matsumoto M, Ishizaki H, Kitazawa M, Aoki J, Fujii D (1995) Cable aerodynamics and its stabilization. In: Proceedings of the ISCD, Liège, pp 289–296
43. Matsumoto M, Shiraishi N, Kitazawa M, Knisely C, Shirato H, Kim Y, Tsujii M (1990) Aerodynamic behaviour of inclined circular cylinders—Cable aerodynamics. J Wind Eng Ind Aerod 33:63–72
44. Matsumoto M, Shirato H, Yagi T, Jones NP, Hayashi T (2001) Field observation system of cable aerodynamics in natural wind. In: Proceedings of the 4th ISCD, Montreal, pp 219–225
45. Matsumoto M, Yagi T, Oishi T, Adachi Y (2005) Effects of axial flow and Kármán vortex interferences on dry-state cable galloping of inclined stay-cables. In: Proceedings of the 6th ISCD, Charleston
46. Matsumoto M, Yagi T, Hatsuda H, Shima T, Tanaka M (2007) Sensitivity of dry-state galloping of cable stayed bridges to Scruton number. In: Proceedings of the ISCD, Vienna, pp 331–338
47. Matsumoto M, Yamagishi M, Aoki J, Shiraishi N (1995) Various mechanism of inclined cable aerodynamics. In: Proceedings of the 9th ICWE, New Delhi, pp 759–770
48. Yoshimura T, Inoue A, Kaji K, Savage MG (1989) A study on the aerodynamic stability of the Aratsu Bridge. In: Proceedings of the Japan-Canada workshop on bridge aerodynamics, Ottawa, pp 41–50

49. Kovács I, Strommen E, Hjorth-Hansen E (1999) Damping devices against cable oscillations on Suuingesund bridge. In: Proceedings of the 3rd ISCD, Trondheim, pp 145–150
50. Langsø H, Larsen O (1987) Generating mechanisms for cable-stay oscillations at the Farø Bridge. In: Proceedings of the international conference on cable-stayed bridges, Bangkok, pp 1023–1033
51. Hikami Y, Shiraishi N (1988) Rain–wind induced vibrations of cables in cable-stayed bridges. J Wind Eng Ind Aerod 29:409–418
52. Kusakabe T, Yokoyama K, Kanazaki T, Sekiya M (1995) The effects of cable cross ties for wind-induced vibration. In: Proceedings of the 9th ICWE, New Delhi, pp 783–792
53. Ruscheweyh H, Verwiebe C (1995) Rain–wind-induced vibrations of steel bars. In: Proceedings of the 1st ISCD, Liège, pp 469–472
54. Flamand O (1994) Rain–wind induced vibration of cables. In: Proceedings of the cable-stayed and suspension bridges, Deauville, 2:523–531
55. Kobayashi H, Minami Y, Miki M (1995) Prevention of rain–wind induced vibration of an inclined cable by surface processing, In: Proceedings of the 9th ICWE, New Delhi, pp 753–758
56. Saito T, Matsumoto M, Kitazawa M (1994) Rain–wind excitation of cables on cable-stayed Higashi-Kobe Bridge and cable vibration control. I In: Proceedings of the cable-stayed and suspension bridges, Deauville, 2:507–514
57. Miyata T, Yamada H, Hojo T (1994) Aerodynamic response of PE stay cables with pattern-indented surface. In: Proceedings of the cable-stayed and suspension bridges, Deauville 2:515–522
58. Irwin PA, Nedim A, Telang N (1999) Wind induced stay cable vibrations – A case study. In: Proceedings of the 3rd ISCD, Trondheim, pp 171–176
59. Yamada H (1997) Control of wind-induced cable vibrations from a viewpoint of the wind resistant design of cable-stayed bridges. In: Proceedings of the 2nd ISCD, Tokyo, pp 129–138
60. Matsumoto M, Yokoyama K, Miyata T, Fujino Y, Yamaguchi H (1989) Wind-induced cable vibration of cable-stayed bridges in Japan. In: Proceedings of the Canada-Japan workshop on bridge aerodynamics, Ottawa, pp 101–110
61. Cheng SH, Tanaka H (2002) Aerodynamics of dry inclined cables. In: Proceedings of the 2nd international conference advances in wind & structures, Pusan, Korea, pp 361–368
62. Cheng SH, Tanaka H, Irwin PA, Jakobsen JB (2003) Aerodynamic instability of inclined cables. In: Proceedings of the 5th ISCD, Santa Margherita, pp 69–76
63. Cheng SH, Irwin PA, Jakobsen JB, Lankin J, Larose GL, Savage MG, Tanaka H, Zurell C (2003) Divergent motion of cables exposed to skewed wind. In: Proceedings of the 5th ISCD, Santa Margherita, pp 271–278
64. Cheng SH, Tanaka H (2005) Correlation of aerodynamic forces on an inclined circular cylinder. Wind Struct 8(2):135–146
65. Larose GL, Zan SJ (2001) The aerodynamic forces on the stay cables of cable-stayed bridges in the critical Reynolds number range. In: Proceedings of the 4th ISCD, Montreal, pp 77–84
66. Macdonald JHG, Larose GL (2006) A unified approach to aerodynamic damping and drag/lift instabilities, and its application to dry inclined cable galloping. J Fluid Struct 22:229–252
67. Larose GL, Jakobsen JB, Savage MG (2003) Wind-tunnel experiments on an inclined and yawed stay-cable model in the critical Reynolds number range. In: Proceedings of the 11th ICWE, Lubbock, Texas 2:1705–1712
68. Larose GL, Jakobsen JB, Savage MG (2003) Wind tunnel experiments on an inclined and yawed circular cylinder in the critical Reynolds number range. In: Proceedings of the 5th ISCD, Santa Margherita, pp 279–286
69. Macdonald JHG (2005) Quasi-steady analysis of inclined cable galloping in the critical Reynolds number range. In: Proceedings of the 6th ISCD, Charleston, South Carolina
70. Ruscheweyh H, Hirsch G (1974) Vibration measurements at the cable stayed Köhlbrand Bridge in Hamburg (in German). Technical Report, Institute for Lightweight Structures, University of Aachen (RWTH)

71. Wianecki J (1979) Cables wind excited vibrations of cable-stayed bridges, In: Proceedings of the 5th ICWE, Fort Collins, Colorado, pp 1381–1393
72. Miyasaka Y, Ohshima K, Nakabayashi S (1987) Experimental study on Ajikawa Bridge cable vibration (in Japanese). Hanshin expressway public corporation engineering report (7)
73. Ohshima K Nanjo M (1987) Aerodynamic stability of the cables of a cable-stayed bridge subjected to rain: a case study of the Aji river bridge. In: Proceedings of the US-Japan joint seminar on natural resources
74. Yoshimura T, Savage MG, Tanaka H (1995) Wind induced vibrations of bridge stay-cables. In: Proceedings of the 1st ISCD, Liège, pp 437–444
75. Lilien JL, Pinto da Costa A (1994) Vibration amplitudes caused by parametric excitation of cable stayer structures. J Sound Vib 174(1):69–90
76. Cremer JM, Counasse C, de Ville de Goyet V, Lothaire A, Dumortier A (1995) The stays, their dynamic behaviour, their equipments: Bridges at Ben-Ahin, Wandre and upon Alzette. In: Proceedings of the 1st ISCD, Liège, pp 489–496
77. Cheng SH (1999) Personal communication
78. Gu M, Liu CJ, Lou GQ, Lin ZX, Xiang HF (1998) Rain–wind-induced vibration of cables on cable-stayed bridges and its control (in Chinese). Shanghai J Mech 25(1):281–288
79. Geurts C, Vrouwenvelder T, Staalduinen P, Reusink J (1998) Numerical modelling of rain–wind-induced vibration: Erasmus bridge, Rotterdam. Struct Eng Int 8(2):129–135
80. Irwin PA (2005) Field monitoring of cable supported bridges. In: Proceedings of the 6th ISCD, Charleston, South Carolina
81. Hardy C, Watts JA, Brunelle J, Cloutier LJ (1975) Research on the dynamics of bundled conductors at the Hydro-Québec Institute of Research. Trans. E+O Div., CEA 14(4)
82. Zasso A, Bocciolone M, Brownjohn J (1992) Rain–wind aeroelastic instability of the inclined hangers of a suspension bridge. In: Proceedings of the inaugural conference of the wind engineering society, Cambridge, UK
83. Main JA, Jones NP, Yamaguchi H (2001) Characterization of rain–wind induced stay-cable vibrations from full-scale measurements. In: Proceedings of the 4th ISCD, Montreal, pp 235–242
84. Narita N, Yokoyama K (1991) A summarized account of damping capacity and measures against wind action in cable-stayed bridges in Japan. In: Ito M, Fujino Y, Miyata T, Narita N (eds) Cable-stayed bridges: recent developments and their future. Elsevier, Amsterdam, pp 257–276
85. Zuo D, Jones NP, Main JA (2004) Vortex- and rain–wind-induced stay cable vibrations in a three-dimensional environment. In: Proceedings of the 5th international colloquium on bluff body aerodynamics & applications, Ottawa, pp 397–400
86. Hjorth-Hansen E, Strømmen E (2003) Wind-excited vibration of the longest suspenders of a suspension bridge. In: Proceedings of the 5th ISCD, St. Margherita, pp 541–543
87. Asa-Jakobsen K, Jordet E, Rambjør SK (1995) Full scale measurements of the behaviour of the Helgeland Bridge, a cable-stayed bridge located in a harsh environment. In: Proceedings of the 1st ISCD, Liège, pp 473–480
88. Andersen H, Hommel D, Veje E (1999) Emergency rehabilitation of the Zárate-Brazo Largo Bridges, Argentina. In: Proceedings of the IABSE conference on cable-stayed bridges—past, present and future, IABSE
89. Caetano ES, Cunha AF, Frias R (2003) Identification of parametric excitation at the international Guadiana bridge. In: Proceedings of the 5th ISCD, St. Margherita, pp 525–532
90. Macdonald JHG, Irwin PA, Fletcher MS (2002) Vortex induced vibrations of the second Severn crossing cable-stayed bridge—Full scale and wind tunnel measurements. Proc ICE: Struct Build 152 (2):123–134
91. Scruton C, Walshe DEJ (1957) A means for avoiding wind-excited oscillations of structures with circular or nearly circular cross-section. NPL Report NPL/Aero/335
92. Macdonald JHG (2001) Susceptibility of inclined bridge cable to large amplitude vibrations considering aerodynamic and structural cable damping. In: proceedings of the 4th ISCD, Montreal, pp 243–250

93. Yamaguchi H, Fujino Y (1998) Stayed cable dynamics and its vibration control. In: Esdahl S, Larsen A (eds) Bridge aerodynamics, Balkema, pp 235–253
94. Electric Power Research Institute (1979) Transmission line reference book: wind-induced conductor motion. EPRI, Palo Alto, Calif
95. Pacheco BM, Fujino Y, Sulekh A (1993) Estimation curve for modal damping in stay cables with viscous damper. J Struct Eng ASCE 119(ST6):1961–1979
96. Krenk S, Høgsberg JR (2003) Optimal damping of cables by an external damper. In: proceedings of the 5th ISCD, Santa Margherita, pp 419–426
97. Hoang N, Fujino Y (2007) Analytical study on bending effect in a stay with a damper. J Eng Mech ASCE 133(11):1241–1246
98. Fujino Y, Hoang N (2008) Design formulas for damping of stay cable with a damper. J Struct Eng ASCE 134(2):269–278
99. Bournand Y, Crigler J (2005) The VSL friction damper for cable-stayed bridges: Some results from maintenance and testing on long cables. In: proceedings of the 6th ISCD, Charleston, South Carolina, pp 199–204
100. Weber F, Distl H, Feltrin G, Motavalli M (2005) Evaluation procedure of decay measurements of a cable with passive-on operating MR damper. In: proceedings of the 6th ISCD, Charleston, South Carolina
101. Miyasaka Y, Nanjo J, Nanjo M, Kado H, Ishitobi T (1992) Cable vibration and its countermeasure of the Tempozan bridge (in Japanese). Bridge Foundation Eng 26(4):27–36
102. Mori Y, Ishitobi T, Nanjo M (1992) Field observation of cable vibration of the Tempozan Bridge and its counter measures(in Japanese). In: proceedings of the 12th national symposium on wind engineering, Tokyo, pp 273–278
103. Fujiwara T, Manabe Y, Yamaguchi K (1999) Vibration control of cables of the Tatara bridge (in Japanese). Bridge Foundation Eng 33(5):16–19
104. Takeguchi M (2000) Aerodynamic stabilization of hanger ropes of the Akashi Kaikyo bridge (in Japanese). Honshi Technical Report BOEA 24(93):18–25
105. Wenzel H (2009) Health monitoring of bridges. John Wiley & Sons, UK

Chapter 11
Vibration Control for Bridge Members

11.1 Introduction

Not only the whole bridge structure but each structural member of a bridge sometimes suffers wind-induced vibrations, causing not necessarily a fatal destruction but other consequences such as fatigue damages. In this chapter, these vibration problems and possible methods to control them are introduced.

11.2 Vibration of Hangers and Their Control

11.2.1 Example of Shitoku Bridge

Figure 11.1 shows the Shitoku Bridge, a Langer-type arch with inclined hangers, having a centre span of 150 m [1, 2]. Hangers of the bridge, up to 22 m long, were made of steel pipes with the outer diameter of 318.5 mm. Aerodynamic excitation of hangers near the span centre was first observed at the wind speed of 5–6 m/s. When the wind speed increased, the vibration of hangers shifted toward both ends of the bridge and even the shortest hanger started vibrating at the wind speed of 15–17 m/s. All through the case, the wind was almost perpendicular to the bridge axis and the hangers vibrated in their plane. Cracks were found developing at more than thirty panel points as a result. The double amplitude and frequency of vibration for one of the central hangers measured at the site are presented in Fig. 11.2. It is obvious that the hangers were excited by vortex shedding in the bending modes. Gusset structures were improved to reduce the stress concentration, and the vortex-induced vibration was effectively suppressed by installing the trip wires around the hangers as shown in Fig. 11.3. Three wires with 32 mm diameter, corresponding to one-tenth of the hanger diameter, were rapped on each hanger with the skew angle of 30°.

Y. Fujino et al., *Wind Resistant Design of Bridges in Japan*,
DOI 10.1007/978-4-431-54046-5_11, © Springer 2012

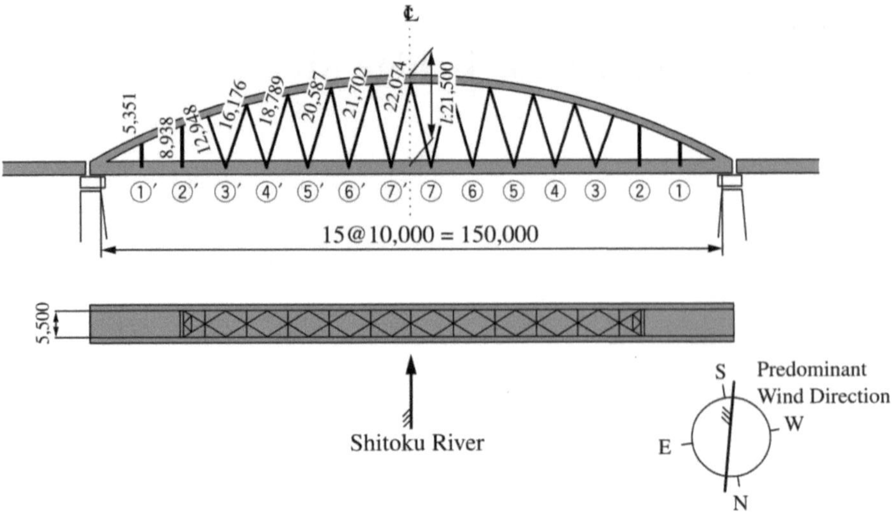

Fig. 11.1 General view of Shitoku Bridge

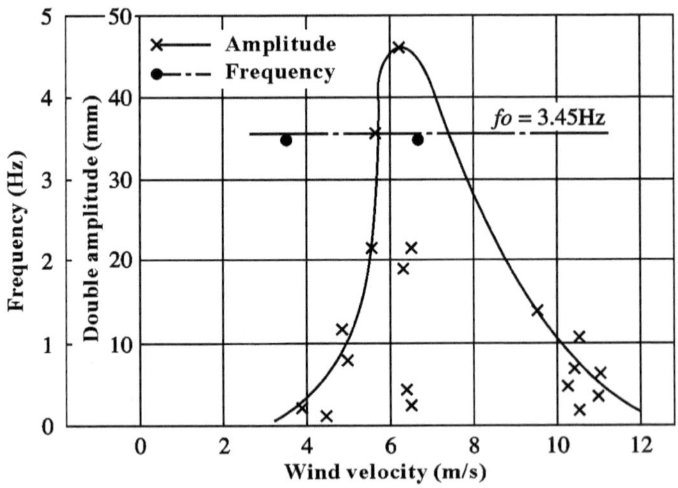

Fig. 11.2 Aerodynamic vibration of central hanger of Shitoku Bridge [2]

11.2.2 Vibration Control Methods in General

As observed in the above example, the tension members with large slenderness ratio sometimes experience the vortex-shedding excitation in bending, or even galloping as it was observed in the Mukaijima Bridge [3]. The wind-induced

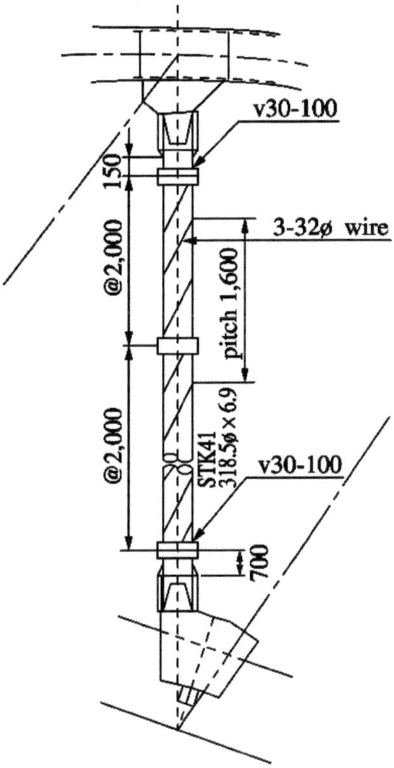

Fig. 11.3 Trip wire of Shitoku Bridge

vibration of bridge members can be controlled by aerodynamic or structural means, or sometimes by the combination of them.

The aerodynamic means that have been adopted in the past include (1) the installation of trip wires, as the case of the Shitoku Bridge, to break up the formation of Kármán vortices; (2) modification of the cross-section by, for example, making holes on a wide-flange section [4]; or (3) cutting out the sharp corners of the section such as a rectangle [5]. The structural means, on the other hand, include the following: (1) installation of struts or connection of tensioned cables between hangers; (2) installation of dampers; (3) reinforcement of hangers; and (4) addition of mass on hangers.

11.3 Vibration Control for Utilities Attached to Bridges

11.3.1 Light Poles

The light poles of the Ohnaruto Bridge, one of the Honshu-Shikoku bridges in the Kobe-Awaji-Naruto route, suffered wind-induced vibration. The cross-section of the pole was hexagonal and was tapered along its height. A newly developed chain

Fig. 11.4 Chain dampers for lighting poles of the Ohnaruto Bridge

damper, shown in Fig. 11.4 [6], was installed for this case. For other light poles with circular cross-sections of the Honshu-Shikoku bridges, the vortex-induced vibration was predicted by wind tunnel tests. The vibration control was intended either by structurally reinforcing the poles or by the installation of the deflectors.

11.3.2 Handrails

The handrails of the Ohnaruto Bridge were also found damaged by the vortex-induced vibration. For this case, the rails were either connected to each other by small flat plates as shown in Fig. 11.5, or by attaching the trip wires as seen in Fig. 11.6.

11.3.3 General Remarks on Aerodynamics of Utilities

Generally speaking, the attached utilities on the bridge deck should be carefully installed so that they are not easily damaged by wind, which may be accelerated by the presence of the bridge deck itself. Sometimes the wind is very smooth without much turbulence but it could be also quite turbulent in the wake of other structures. It has been reported that some of the light poles of the cable-stayed bridges were damaged due to vibration induced in the wake of the bridge towers.

Fig. 11.5 Connections of handrails with flat plates, at the Ohnaruto Bridge

Fig. 11.6 Trip wires on handrails at the Ohnaruto Bridge

References

1. Japan Road Association (1991) Wind resistant design manual for highway bridges (in Japanese). Maruzen, Tokyo. Also revised version in 2007
2. Narita N (1971) Vibrations of bridge members due to winds (in Japanese). Bridge Foundation Eng 5(9):1–5
3. Makihata T, Ushio M, Izuno H (1972) Wind-resistance study of hanger of Mukaijima Ohashi bridge (in Japanese). J Bridges 8(5):84–96

4. Agawa T, Ikenouchi M, Inoue H (1980) A study on aerodynamic instability of bridge structure member with rectangular cross section (in Japanese). In: Proceedings of the 6th national symposium on wind engineering, Tokyo, pp 325–332
5. Nanjo M, Ushio M (1990) Wind tunnel experiment and field observation on the rectangular cylinders with square corner cuts (in Japanese). In: Proceedings of the 11th national symposium on wind engineering, Tokyo, pp 323–329
6. Yamada K, Akimoto S (1985) Wind proof study of illumination post of the Ohnaruto Bridge (in Japanese). Honshi Technical Report, BOEA (36):26–29

Chapter 12
Wind Effects on Traffic

12.1 Introduction

In this chapter, a brief description of adverse wind effects on traffic and pedestrians and their control methods are given.

12.2 Protection of Traffic from Adverse Wind Effects

12.2.1 *Example of the Ohnaruto Bridge*

It has been understood that a sudden change of wind speed and/or its direction on traveling vehicles is one of the major causes of traffic accidents on highway bridges [1]. In case of suspension bridges, the existence of large towers and anchorages often distort the wind conditions on traffic. Figure 12.1 shows an example of the Severn Bridge, UK, where the wind shields were installed near the towers so that the change of wind conditions on traffic would become more tolerable. Similar attempts were made for the Ohnaruto Bridge as seen in Fig. 12.2. There have been many research works, including both wind tunnel tests and field measurements, regarding the wind properties behind the shields and their influence on vehicle maneuverability [2–5].

12.2.2 *Example of the Bannosu Viaduct*

The Bannosu Viaduct (Fig. 12.3), a double-deck viaduct of the Kojima-Sakaide route of the Honshu-Shikoku Bridges, is supported by tall, massive concrete piers of 50–70 m high and 5–10 m wide. The local wind at the bridge site is, as a result, considerably reduced but significantly turbulent near the piers and there exists large

Y. Fujino et al., *Wind Resistant Design of Bridges in Japan*, DOI 10.1007/978-4-431-54046-5_12, © Springer 2012

Fig. 12.1 Wind shields around the tower of the Severn Bridge

Fig. 12.2 Wind shields around the tower of the Ohnaruto Bridge

wind speed difference from the center span location. The wind shields as shown in Fig. 12.4 have been adopted in order to mitigate the adverse wind effects on traffic for this case [6].

12.2.3 Example of Yoshinogawa Ohashi Bridge

The Yoshinogawa Ohashi Bridge over the Yoshino river connects the northern and southern parts of the City of Tokushima with 1,137 m bridge total length. The site

Fig. 12.3 Perspective of the Bannosu Viaduct

Fig. 12.4 Wind shields of the Bannosu Viaduct

of the bridge is about 3 km away from the seacoast and strong winds due to typhoons and seasonal monsoon have been frequently observed. At the requests of the local cyclists and pedestrians, the polycarbonate shielding plates were attached to the handrails as shown in Fig. 12.5. It should be pointed out, however, that this kind of modification could significantly alter the aerodynamic characteristics of the bridge deck itself and hence needs to be critically investigated.

Fig. 12.5 Wind shields of the Yoshinogawa Ohashi Bridge

References

1. Utsunomiya H, Nagao F, Yano Y (1985) Strong wind properties around Ohnaruto Bridge (in Japanese). In: Proceedings of the 40th JSCE annual meeting 1:413–414
2. Okajima A (1986) Aerodynamic forces acting on a vehicle passing in the wake of a bridge tower (in Japanese). In: Proceedings of the 41th JSCE annual meeting 1:635–636
3. Charuvisit S, Kimura K, Fujino Y (2004) Effects of wind barrier on a vehicle passing in the wake of a bridge tower in cross wind and its response. J Wind Eng Ind Aerod 92:609–639
4. Charuvisit S, Kimura K, Fujino Y (2004) Experimental and semi-analytical studies on the aerodynamic forces acting on a vehicle passing through the wake of a bridge tower in cross wind. J Wind Eng Ind Aerod 92:749–780
5. Smith BW, Baker CP (1998) Design of wind screens to bridges. In: Larsen A, Esdahl S (eds) Bridge aerodynamics. Balkema, Copenhagen, Denmark, pp 289–298
6. Tatsumi M, Yasuda M, Matsumoto T (1999) Inspections and repair works of the Seto Ohashi bridge for ten years after its opening. In: Miyata T, Fujisawa N, Yamada H (eds) Long span bridges and aerodynamics. Springer, Tokyo, pp 81–104

Appendix 1: Wind Tunnel Test Specification for Honshu-Shikoku Bridges (2001)

Wind Tunnel Test Specification for Honshu-Shikoku Bridges (1980) described the wind tunnel testing methods, particularly on the sectional model tests and tower models. This specification was applied to the design of the Seto-Ohashi Bridges, whose maximum span length was 1100 m. When the *Specification for Akashi-Kaikyo Bridge* (1990) (see Sect. 3.2) was developed, it further included dynamic tests of the towers after the bridge's completion and taut-strip model tests, supported by the updated knowledge of wind engineering.

However, during the design of the Akashi-Kaikyo Bridge, the importance of more detailed investigation toward three-dimensionality of both wind and structure was recognized, in order to avoid overlooking unknown phenomena that may occur due to its unprecedented nearly 2000 m long span. Therefore, the aeroelastic full bridge model tests and flutter analysis were conducted. The flutter analysis was used to supplement what the full bridge model test could not cover due to limited test cases and so on. A large boundary layer wind tunnel facility was newly constructed in Tsukuba City for this purpose in 1991. After Akashi, the aeroelastic full bridge model tests of the Tatara and Kurushima-Kaikyo bridges were also carried out in the same wind tunnel. Since there was a strong possibility that the wind-induced behavior of these bridges would be significantly influenced by the local topography of the bridge site, it was advantageous to perform aeroelastic full bridge model tests with properly scaled terrain models in a large wind tunnel.

In order to summarize the testing methods that were developed during the design stage of the Akash-Kaikyo Bridge, Kurushima-Kaikyo Bridges and Tatara Bridge, *Wind Tunnel Test Specification for HSB* (2001) was formulated. It specifies various wind tunnel testing methods, particularly the methods of the aeroelastic full bridge tests and measurement of unsteady aerodynamic forces needed for numerical analyses. It should be noted that contents of the document were edited from the original document written in Japanese for the convenience of foreign readers.

In the followings, some items are extracted from the Specifications.

Y. Fujino et al., *Wind Resistant Design of Bridges in Japan*,
DOI 10.1007/978-4-431-54046-5, © Springer 2012

A.1 Wind Tunnel Test Specifications for Honshu-Shikoku Bridges (2001)

A.1.1 Scope

This specification describes the testing methods applicable to the aeroelastic full bridge model tests and the measurement of unsteady aerodynamic force coefficients. Contents of this document were developed in the course of the wind resistant design of the Akashi-Kaikyo, Tatara and Kurushima-Kaikyo bridges. The reader should refer to the *Wind Tunnel Test Specification for Akashi Kaikyo Bridge* (1990) for information regarding the measurement of static force components and the dynamic tests of sectional deck models and tower models.

A.1.2 Aeroelastic Full-Bridge Model Test

A.1.2.1 Damping

Standard structural damping (in logarithmic decrement) of an aeroelastic full bridge model at no wind condition is shown in Table A1.1.

A.1.2.2 Flow Characteristics

General

The air flow at the wind tunnel test section should be uniform and smooth, unless turbulence is intended to be created such as the simulated natural wind turbulence. When a boundary layer simulation of natural wind is introduced, the turbulence intensity of mean flow direction component is specified to be 10% at the bridge deck elevation, and the vertical profile of the mean wind speed is specified in the *Standard* (Table 3.3). Also a weakly turbulent flow with 5% intensity at the bridge deck elevation is to be used for evaluation of turbulence effects on vortex-induced vibration. These values of turbulence intensity and wind speed profile are based on the fact that many of HSBs are located at open terrain above the sea.

Simulated Turbulence Generation and Its Evaluation

- When terrain models are not used, the boundary layer flow should have a designated vertical profile of the mean wind speed and turbulent intensity at the bridge deck elevation generated by appropriate turbulence generator(s).

Table A1.1 Logarithmic decrement in design value

Mode and structure		Bending vibration		Torsional vibration	
		Logarithmic decrement	Reference amplitude	Logarithmic decrement	Reference amplitude
Deck-vibration dominant mode	Stiffening truss	0.03	1/200 of chord	0.02	0.5°at representative point
	Closed-box deck	0.02		0.02	
Tower-vibration dominant mode	Tower-cable-deck system	0.02	1/500 of completed tower height	0.02	
	Free-standing tower	0.01		0.01	

- When terrain models are used, the approaching boundary layer flow should simulate the characteristics of turbulence at the bridge deck elevation as close as possible.
- The simulation of turbulence characteristics should be evaluated by the power spectral density function of the longitudinal and vertical components of wind velocity, the turbulence intensities at the bridge deck elevation, and the vertical profiles of mean wind speed.

A.1.2.3 Measurement

Vertical Inclination Angle of Approaching Flow

In smooth flow, aeroelastic full bridge model tests should be conducted for a range of vertical inclination angle of approaching flow from $-3°$ to $+3°$. On the other hand, the tests should be conducted for $0°$ inclination angle in boundary layer turbulence.

Note: It is often experienced that simulation of the target boundary layer turbulence in the wind tunnel is difficult. Besides, it is specified that verification of divergent type vibration by section model test should be conducted in a smooth flow with angle of attack from $-3°$ to $+3°$. In this regard, the basic idea is to conduct an aeroelastic model test in a smooth flow as well as boundary layer turbulence. Vertical inclination angle of approaching flow for an aeroelastic model test is realized by means of slope panels and/or model inclination. In the case of positive angle of inclination (upward direction), slope panels are often used by setting them under the model. The model inclination method is used for both positive and negative inclination angle. The slope panel method is better because the gravity force effect is not exerted on the model. Some discussions on the setup of angle of attack are given in Sect. 6.4.2.

Range of Test Wind Speed

The tests should be conducted at least up to the reference wind speed in model scale. The wind speed increment should be properly chosen so that the response characteristics of the bridge would be identified as clearly as possible.

Range of Test Amplitude

For the response measurement, sensors should be appropriately allocated so that the mode shapes can be identified. In order to clarify the vibration characteristics, it is desirable to continue the measurement until the torsional amplitude of $5°$ and/or the bending amplitude of 1/20 of the chord length is reached. It is also important to clarify the vibration characteristics for each mode.

A.1.3 *Measurement of Unsteady Aerodynamic Forces*

A.1.3.1 General

This section describes measurement of unsteady aerodynamic forces which are used for flutter analysis. There are, in general, three kinds of method of unsteady aerodynamic force measurement; free vibration method, forced vibration method and method by integration of unsteady surface pressure (Sect. 6.2.6). Forced vibration method directly detects unsteady aerodynamic forces by vibrating the model in airflow. Principle of the measurement is straightforward; however special attention must be paid to the removal of inertia force of the moving model which will be described later. In the *Specification*, measurement of unsteady aerodynamic forces by forced vibration method is described.

A.1.3.2 Model Support

Since the forced vibration method excites the model in airflow, the inertia force and unsteady aerodynamic forces are detected. The inertia force is much bigger than the aerodynamic forces. Therefore, accurate subtraction of the inertia force is the key to the high precision measurement. The inertia force can be removed by subtracting it after the measurement or balancing it with counterweights in the measurement that directly gives only the aerodynamic forces.

In the case of the Akashi-Kaikyo Bridge, the importance of unsteady forces with along-wind movement of the deck and unsteady aerodynamic drag was pointed out. However, facilities which could measure such unsteady aerodynamic forces were not common, and those forces may not be important for shorter span or other stiffening girder types. Therefore, minimum requirement of the measurements for lift and pitching moment components with vertical and torsional motion is specified in the *Specification*.

A.1.3.3 Measurement

The model driving amplitude is to be 1% of the model chord length in bending motion, and 1° in torsion. These amplitudes correspond to the criteria of onset wind speed for divergent type vibration.

Table A1.2 Dimension of bridge deck model

Type of test section \ Items	Chord length/effective height of test section	Model length/chord length	Blockage ratio
Closed type	≤ 0.4	≥ 2	$\leq 5\%$
Open type	≤ 0.2	≥ 3	$\leq 5\%$

A.2 Wind Tunnel Test Specifications for Akashi-Kaikyo Bridge (1990)

A.2.1 Common Rules

A.2.1.1 Scope

This specification shall be applied to wind tunnel test (excluding aeroelastic full-bridge model test in large wind tunnel) of the Akashi-Kaikyo Bridge.

A.2.1.2 Bridge Deck Model

The model dimensions shall be in accordance with Table A1.2.

A.2.1.3 Main Tower Model

Scale Ratio

The linear scale of the model shall be chosen so that the model can represent the structural details of the prototype with sufficient accuracy. The minimum scale ratio is to be 1/250 and 1/500 for the tests with smooth flow and shear flow turbulence, respectively.

Note: The scale of 1/250 in smooth flow is restricted by requirement of accurate reproduction of model details. On the other hand, the scale of 1/500 in shear flow is restricted by shear flow height in wind tunnel, though accuracy of model details decreases.

Dimensions

The model dimensions shall be in accordance with Table A1.3.

Table A1.3 Dimension of tower model

Height of model/effective height of test section	Distance between shafts/effective width of test section	Blockage ratio
≤0.8	≤0.2	≤5%

A.2.2 Measurement of Aerodynamic Forces

A.2.2.1 Measurement

Vertical Angle of Attack and Horizontal Azimuth Angle

For the force measurement, the vertical angle of attack should be altered from −15° to +15° with 1° increment, and the horizontal yaw angle from 0° to 75° for every 15°.

Wind Speed

The force measurement should be conducted at two different wind speeds, with 10 and 20 m/s as standard.

Precision Required for the Tests

The error in model setting should be less than 0.1° in vertical angle of attack and 0.5° in yaw angle. For each set of conditions, the force measurement should be repeated more than twice. If the difference between measurements is more than 0.05 in drag and side force coefficients or more than 0.01 in lift force and pitching moment coefficients, the test should be further repeated to determine reliable data.

A.2.3 Section Model Tests

A.2.3.1 Measurement

Vertical Angle of Attack

Response measurement of bridge deck section model should generally be conducted at every 1° from −3° to +3° attack of angle. Change of the attack angle caused by the torsional deformation due to the wind load should be adjusted at each wind speed case, if necessarily.

Table A1.4 Tolerance in the setting of section model test

Item	Mass	Polar moment of inertia	Frequency ratio	Logarithmic decrement
Allowable deviation	±2%	±2%	±5%	±0.005

Table A1.5 Tolerance in the setting of section model test

Items	Torsional displacement	Bending displacement
Accuracy	±0.05°	Chord length × 1/2000

Range of Test Wind Speed

The test should generally be conducted in wind speed up to 1.2 times of the design wind speed. Interval of test wind speed should properly be chosen so that the detail characteristics of the phenomena in concern can be clarified.

Precision Required for the Tests

The tolerance in model setting is given in Table A1.4. Accuracy of the measurement in the response measurement of the section model test is given in Table A1.5.

Appendix 2: Bridge Map in Japan

The locations of bridges referred in this book are shown in the following Figures A2.1–A2.3. Each prefecture is numbered from north to the south. The bridge names in the figures are classified by this number and listed in Table A2.1 together with the name of prefecture. In Fig. A2.1, bridge locations are shown in a map of Japan. Detailed locations are shown in Fig. A2.2 for bridges in Tokyo and Osaka areas. Detailed locations of Honshu-Shikoku and nearby bridges are shown in Fig. A2.3.

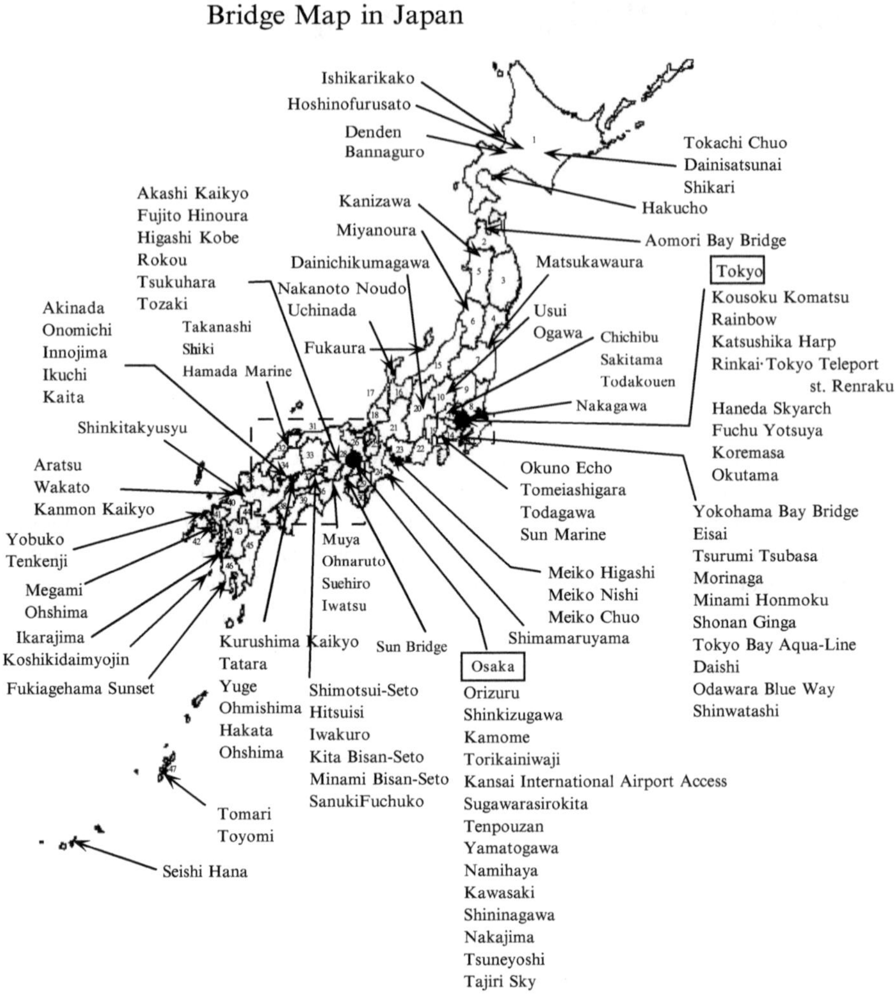

Fig. A2.1 Bridge map in Japan

13. Tokyo	14. Kanagawa
Kousokukomatsu	Yokohama Bay Bridge
Rainbow	Tsurumi Tsubasa
Katsushika Harp	Morinaga
Rinkai・Tokyo Teleport st. Renraku	Minami Honmoku
Haneda Sky Arch	Trans-Tokyo Bay Highway
Fuchu Yotsuya	Daishi
Koremasa	Odawara Blue Way
Okutama	

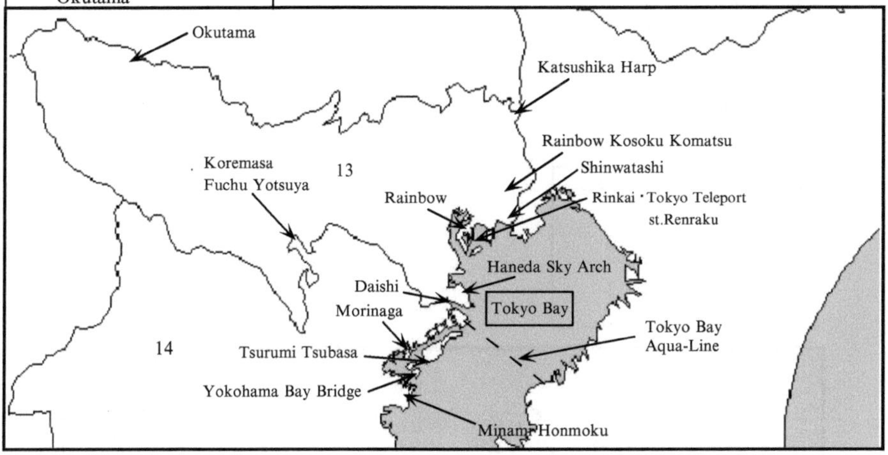

27. Osaka
Orizuru
Shinkizugawa
Kamome
Torigainiwaji
Kansai International Airport Access
Sugawarasirokita
Tenpouzan
Yamatogawa
Namihaya
Kawasaki
ShinInagawa
Nakajima
Tsuneyoshi Renraku
Tajiri Sky

28. Hyogo
Akashi Kaikyo
Higashi Kobe
Rokkou
Tsukuhara
Tozaki

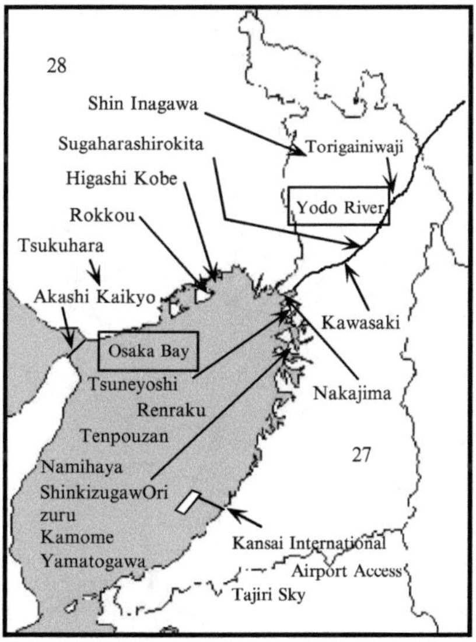

Fig. A2.2 Bridge map in Tokyo and Osaka

28.	Hyogo
	Akashi-Kaikyo
34.	Hiroshima
	Akinada
	Shin Onomichi
	Innoshima
	Ikuchi
	Kaita
36.	Tokushima
	Ohnaruto
37.	Kagawa
	Shimotsui-Seto
	Hitsuishijima
	Iwakurojima
	Kita Bisan-Seto
	Minami Bisan-Seto
	Sanuki Fuchuko
38.	Ehime
	Kurushima-Kaikyo
	Tatara
	Yuge
	Ohmishima
	Hakata
	Ohshima

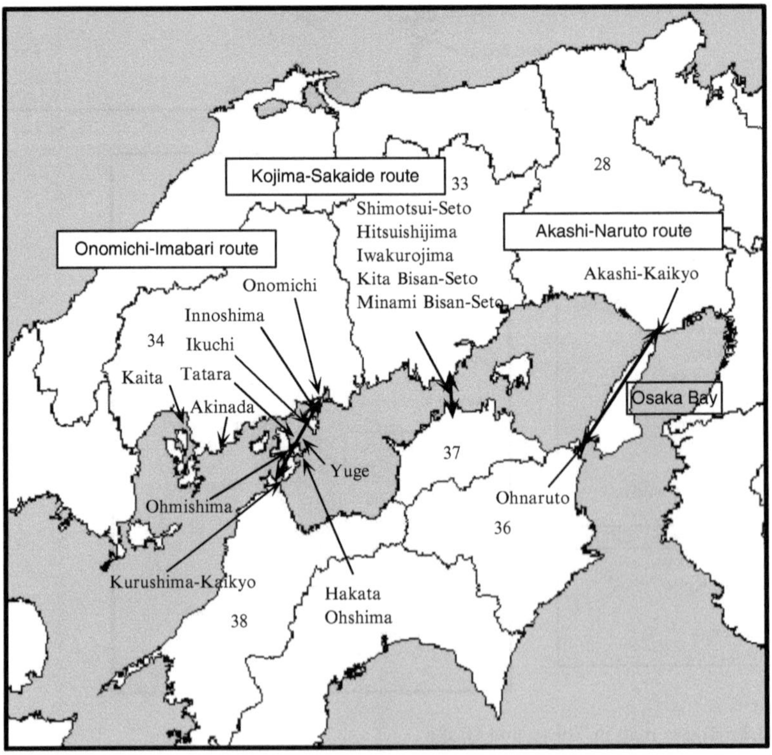

Fig. A2.3 Map of Honshu-Shikoku and nearby bridges

Table A2.1 List of bridges in the following maps

1. *Hokkaido*
   - Hakucho
   - Tokachi Chuo
   - Daini Sat sunai
   - Ishikarikako
   - Hoshinofurusato
   - Denden
   - Banna guro
   - Shikari

2. *Aomori*
   - Aomori Bay Bridge

5. *Akita*
   - Kanizawa

6. *Yamagata*
   - Miyanoura

7. *Fukushima*
   - Matsukawaura

8. *Ibaragi*
   - Todokouen
   - Nakagawa

10. *Gunma*
    - Usui
    - Ogawa

11. *Saitama*
    - Chichibu Park
    - Sakitama

13. *Tokyo*
    - Kousoku Komatsu
    - Rainbow
    - Katsushika Harp
    - Rinkai . Tokyo Teleport st. Renraku
    - Haneda Sky Arch
    - Fuchu Yotsuya
    - Koremasa
    - Okutama
    - Shinwatashi

14. *Kanagawa*
    - Yokohama Bay Bridge
    - Eisai
    - Tsurumi Tsubasa
    - Morinaga
    - Minami Honmoku
    - Shonan Ginga
    - Tokyo Bay Aqua-Line
    - Daishi
    - Oda wara Blue Way

15. *Niigata*
    - Fukaura

17. *Ishikawa*
    - Nakanoto Noudo
    - U china da

20. *Nagano*
    - Daini Chikumagawa

22. *Shizuoka*
    - Okuno Echo
    - Tomei Ashigara
    - Todagawa
    - Sun Marine

23. *Aichi*
    - Meiko Higashi
    - Meiko Nishi
    - Meiko Chuo

24. *Mie*
    - Shimamaruyama

27. *Osaka*
    - Orizuru
    - Shinkizugawa
    - Kamome
    - Torikainiwaji
    - Kansai International Airport Access
    - Sugawarasirokita
    - Tenpouzan
    - Yamatogawa
    - Namihaya
    - Kawasaki
    - Shin Inagawa
    - Nakajima
    - Tsuneyoshi Renraku
    - Tajiri Sky

28. *Hyogo*
    - Akashi-Kaikyo
    - Fujito Hinoura
    - Higashi Kobe
    - Rokkou
    - Tsukuhara
    - Tozaki

30. *Wakayama*
    - Sun Bridge

32. *Shimane*
    - Takanashi
    - Shiki
    - Hamada Marine

34. *Hiroshima*
    - Akinada
    - Shin Onomichi
    - Innoshima
    - Ikuchi
    - Kaita

36. *Tokushima*
    - Muya
    - Ohnaruto
    - Suehiro
    - Iwazu

37. *Kagawa*
    - Shimotsui-Seto
    - Hitsuishijima
    - Iwakurojima
    - Kita Bisan-Seto
    - Minami Bisan-Seto
    - Sanuki Fuchuko

38. *Ehime*
    - Kurushima-Kaikyo
    - Tatara
    - Yuge
    - Ohmishima
    - Hakata
    - Ohshima

40. *Fukuoka*
    - Shinkitakyushu
    - Aratsu
    - Wakato
    - Kanmon Kaikyo

41. *Saga*
    - Yobuko
    - Tenkenjibashi

42. *Nagasaki*
    - Megami
    - Ohshima

46. *Kagoshima*
    - Ikara
    - Koshiki Daimyojin
    - Fukiagehama Sunset

47. *Okinawa*
    - Tomari
    - Toyomi
    - Seishi Hana

Index

MIX
Papier aus verantwortungsvollen Quellen
Paper from responsible sources
FSC® C105338

If you have any concerns about our products,
you can contact us on
ProductSafety@springernature.com

In case Publisher is established outside the EU,
the EU authorized representative is:
Springer Nature Customer Service Center GmbH
Europaplatz 3, 69115 Heidelberg, Germany

Printed by Libri Plureos GmbH
in Hamburg, Germany